UNDERSTANDING EVOLUTION

*What every Christian Parent Should Know
and Share with their Family*

Exposing the Fallacies of Darwinism
Through Science

W. A. GURBA

PRESS

UNDERSTANDING EVOLUTION
What every Christian Parent Should Know and Share with their Family
by W. A. GURBA

Cover Photo: Suggests God in the act of Creation

Cover Design: S. Gurba

Printed in the United States of America.

ISBN 9781498470452

www.xulonpress.com

ACKNOWLEDGMENTS

The first and foremost acknowledgment goes gratefully to the Creator. I thank Him for ears to hear and eyes to see His presence in the natural world.

Many an author has given high praise to his wife for her contribution toward making his book a reality. I have read many of these acknowledgments. I now realize, I had no concept of the enormity of that help. It is an understatement to say this book would not be in print except for the input of my beloved wife, Sally. Her assistance and patience throughout this project was of God. I thank her.

DEDICATION

Years of experience in front of the classroom have compelled me to share the need for discernment when reading texts related to science and especially evolution.

This book is dedicated to all the students I have had the privilege of teaching; especially those unsuspecting minds upon whom I may have pushed the tenants of Darwinism, from them I seek forgiveness.

I also dedicate this book to those students with whom I shared the pleasure of discovering God's hand in the natural world.

Lastly it is dedicated to all future students, may they learn, grow and share in an understanding and appreciation for the Creator and His creation.

CONTENTS

Chapter One

INTRODUCTION

"Often a cold shudder has run through me, and I have asked myself whether I may have not devoted myself to a phantasy."*Charles Darwin (1)*

Darwin was right. Darwinian evolution is fantasy masquerading as fact. It is easy to say that now, but the truth is, I not only believed the fantasy, I taught it.

I now realize Darwinian evolution is the centerpiece of atheism.

Did life begin by some random spontaneous occurrence, thus negating the need for a Creator?

Did this original life form, unguided, produce all life as we know it, again negating the need for a Creator?

This is a great theory if you are an atheist, or in the words of Oxford professor Richard Dawkins,

"Darwin made it possible to be an intellectually fulfilled atheist."(2)

Is Darwinian evolution really intellectually fulfilling? Dawkins wrote the above words in 1986. Our knowledge of life processes is far more advanced today compared to what was in 1986. What seemed obvious to Dawkins in 1986 is far from obvious today.

Twenty first century science has exposed Dawkins' basic premise, the power of mutation and cumulative selection, to be incapable of producing the many varieties of complex body plans in existence.

Thus, the basic tenants of Darwinian evolution are being undermined by twenty first century science. Today it is no longer intellectually fulfilling. It is fast becoming an outdated scientific dogma atheists are scrambling to keep alive.

The general public and the Christian community have allowed this unsubstantiated and weak theory to enjoy the status of scientific 'fact' for far too long.

This book takes a serious look at Darwinian evolution. You can decide for yourself if the Darwinist claims are valid or if there is a clear trend in science toward realizing our universe and life in it are products of design.

I taught science for thirty years. Roughly half that time I was not a Christian. I was ambivalent about God. I thought He might be out there but I did not give Him much thought. He was certainly not a part of my everyday life.

The Bible was to me a history book with a lot of fairy tale stories mixed in. My dominant worldview was Darwinian. Being a Chemistry and Biology teacher gave me the perfect platform to push Darwinian evolutionary views. I took every opportunity to indoctrinate my students with the Darwinian worldview while discounting the Bible.

Growing up I had minimal exposure to Christianity. College years were spent studying science. After college I travelled to Australia where I spent five years teaching high school Chemistry and Biology. The next six years were spent in Papua New Guinea teaching and working on the design and implementation of primary school science curriculum.

One of my duties in Papua New Guinea was writing upgrade courses for primary school science teachers. I travelled to different districts throughout New Guinea teaching these upgrades to the teachers. Although Darwinian evolution was in no way connected to any of these courses, I always managed to include it.

A particular occasion comes to mind that I would like to share. During an upgrade course, one of the primary teachers/students, a young man in his twenties, became very interested in the evolution discussion. He stayed after class and we talked evolution for some time. That evening I felt pretty good about making a convert to evolutionary thought.

Meanwhile that young man had gone back to his village and discussed evolution with the village Pastor. The next day this same primary teacher came to class dressed in sackcloth and covered with ashes, the traditional Biblical garments of repentance. Needless to say we no longer discussed evolution.

The above story illustrates the conscious and deliberate steps I took to not only undermine the Biblical creation story but to replace it with Darwinian evolution.

My next teaching position was in St. Petersburg, Florida. I continued to indoctrinate students with Darwinian evolutionary thought whenever the opportunity presented itself.

While in St. Petersburg I began to develop a relationship with God. In 1990 I made a personal commitment of faith to Jesus Christ. After I accepted Jesus Christ as my Savior the scales began to fall from my eyes. I began to see the Darwinian evolution material I was reading and teaching, in a new light.

The Bible in a number of places describes people as having eyes but being unable to see and having ears but being unable to hear. I know from personal experience how true these words are. Once my eyes and ears were opened, Darwinian evolution's words, phrases and concepts began to take on new meanings. I began to see discrepancies, distortions and even misleading ideas. I came to question some of Darwinian evolution's most basic tenants.

Knowing full well I had spent considerable time leading high school students away from the Bible. I decided it was time to start leading them to the Truth.

Over the past fifteen years I developed an eight unit course of one hour presentations. Students are given a factual look at the evolution story. This course provides the information needed to resist the Darwinian worldview.

I have taught this course to youth groups and adult general meetings. This book is an expansion of that course. This book will help you see Darwinian evolution from a new perspective. The purpose of writing this book is to shine scientific light on Darwinian evolution and more specifically to show that Darwinian evolution is far from scientific fact.

When God is acknowledged as Creator, science can undertake the fascinating discovery of unlocking the secrets of His creation.

Now that the book is complete, I also see it as my story. You probably understand the concept of 'having a ministry.' This book is part of my ministry. It is a result of my transition from believing in Darwinian evolution to understanding God as Creator.

It is my hope you will read this book with your family and together find it informative, interesting, and a Biblical anchor in a hostile secular sea.

Chapter Two

THE IDEOLOGICAL STRUGGLE FOR THE HEARTS AND MINDS OF OUR CHILDREN

"Education is the most powerful ally of Humanism, and every American public school is a school of Humanism. What can the theistic Sunday school, meeting for an hour once a week, and teaching only a fraction of the children, do to stem the tide of a five day program of humanistic teaching?"

Charles F. Potter (3)

There is an ongoing struggle for the hearts and minds of our children. Some of my readers might think I am over-stating the problem; perhaps things are not that bad. So, before we continue, let's take a look at a few quotes from those involved in setting the parameters of our children's education.

We begin with a quote from professed atheist J. Dunphy.

"I am convinced that the battle for humankind's future must be waged and won in the public school classroom by teachers who correctly perceive their role as proselytizers of a new faith, a religion of humanity that recognizes and respects the spark of what theologians call divinity in every human being. These teachers must embody the same selfless dedication as the most rabid fundamentalist preachers, whatever subject they teach, regardless of the educational level-preschool day care or large state university. It will undoubtedly be a long, arduous, painful struggle replete with much sorrow and many tears, but humanism will emerge triumphant. It must if the family of humankind is to survive."(4)

Wow! I bet you thought your children went to school to learn the three R's!

Daniel Dennett is a professor at the Center for Cognitive Studies at Tufts University. He gives this warning to parents who teach their children that man is not a product of Darwinian evolution.

"If you want to teach your children that they are the tools of God, you had better not teach them that they are God's rifles, or we will have to stand firmly opposed to you: your doctrine has no glory, no special rights, and no intrinsic and inalienable merit. If you insist on teaching your children false-hoods—that the Earth is flat, that "Man" is not a product of evolution by natural selection—then you must expect, at the very least, that those of us who have freedom of speech will feel free to describe your teachings as the spreading of falsehoods, and will attempt to demonstrate this to your children at our earliest opportunity. Our future well-being—the well-being of all of us on the planet—depends on the education of our descendants."(5)

It is worth noticing the dishonesty Dennett displays when he attempts to associate the teaching of a flat earth theory with the questioning of Darwinian evolution. It is so blatant he should be ashamed of himself. The truth is, the Bible taught the earth was spherical centuries before science was able to prove it. (See 'Science in the Bible' in Chapter 12). Dennett's attempt to equate anyone who questions Darwinian evolution with someone who believes in a flat earth theory is not only nonsense it is deliberately misleading.

However, Dennett's dishonesty serves an important purpose. It shows that discussions about God as Creator versus Darwinian evolution are more ideological than scientific. There are many examples of this throughout the book.

Harvard professor Chester Pierce had this to say at an educational seminar in Denver.

"Every child in America entering school at the age of five is mentally ill because he comes to school with certain allegiance toward our founding fathers, toward his parents, toward our elected officials, toward a belief in a supernatural being, and toward the sovereignty of this nation as a separate entity. It's up to you teachers, to make all of these sick children well by creating the international child of the future. "(6)

Take a minute and turn to appendix A for quotes similar to the above.

Periodically you will be asked to refer to an appendix for additional information. The quotes and information in these appendices are well worth pursuing, I hope you will avail yourself of them.

It is naive to think these quotes are the exception. There is a very real struggle going on for the hearts and minds of our children. This struggle can be found in every aspect of education. One particular discipline is the focal point.

Physicist Steven Weinberg tells us,

"I personally feel that the teaching of modern science is corrosive of religious belief, and I am all for that."(7)

The above statement is an ideological statement not a scientific one. Science is not for or against anything. Science follows the data regardless of where it leads. Weinberg is clearly telling us he hopes to use the **teaching** of science to corrode religious beliefs. He wants science to be taught from the ideological viewpoint he favors, a viewpoint detrimental to religious belief. It is the **teaching** of science that is the tool atheists use to attack religion. The above quote is a clear example showing the struggle is ideological not scientific.

Science is not corrosive to religious belief. There is nothing inherent in science that makes religious belief impossible or irrelevant. The material studied in almost every branch of science does not detract in any way from the Bible or religious belief.

Pathologist and Professor of Medicine, Raul O. Leguizamon, makes this comment:

"I am absolutely convinced of the lack of true scientific evidence in favor of Darwinian dogma. Nobody in the biological sciences, medicine included, needs Darwinism at all. Darwinism is certainly needed, however, in order to pose as a philosopher, since it is primarily a worldview. And an awful one, as George Bernard Shaw used to say."(8)

The content, the actual material of almost all science, is utterly and completely neutral with respect to religious belief. To what then is Weinberg referring? He is referring to only two aspects of science neither of which allow for the possibility of a Creator.

The first of these is the Darwinian Theory that life is the result of purely natural phenomenon. That life began by blind random chance and over time eventually produced humans. In other words life has nothing to do with a Creator.

The second is the ideological bias atheists have put on the study of science. This ideological bias demands everything be studied from only a materialistic viewpoint. Scientific materialism has nothing to do with accumulating worldly goods.

Scientific materialism is a viewpoint that demands anything and everything be explained only by natural causes. The key word in the preceding sentence is the word *only*. It is a highly dogmatic word which dismisses any notion, idea or data suggesting there is anything more to life than mere atoms and molecules. No matter what phenomenon occurs scientific materialism requires it must be explained only by the interaction of atoms and molecules.

This automatically eliminates anything spiritual, fitting nicely with the atheistic viewpoint of the universe. There is absolutely no solid reason to shackle scientific study by limiting it to ONLY atoms and molecules. The sole reason for doing this is to accommodate the atheistic bias in science.

It is rather ironic that atheistic scientists would staunchly defend the idea that they are searching for the truth and yet have unequivocally decided they know the truth in this instance.

Lest there be any doubt as to the ideological viewpoint atheists have put on the study of science, evolutionary biologist Richard Lewontin makes their position very clear when he states:

"Our willingness to accept scientific claims that are against common sense is the key to an understanding of the real struggle between science and the supernatural. We take the side of science in spite of the patent absurdity of some of its constructs, in spite of its failure to fulfill many of its extravagant promises of health and life, in spite of the tolerance of the scientific community for unsubstantiated just-so stories, because we have an a priori, a commitment to materialism. **It is not that the methods and institutions of science somehow compel us to accept a material explanation of the phenomenal world,** but, on the contrary, that we are forced by our a priori adherence to material causes to create an apparatus of investigation and a set of concepts that produce material explanations, no matter how counter-intuitive, no matter how mystifying to the uninitiated. Moreover, that *materialism is absolute*, for we cannot allow a Divine Foot in the door. "(9) (Bold italics are mine.)

When Lewontin states 'it is not that the methods and institutions of science somehow compel us to accept a material explanation of the phenomenal world', just what is he saying?

He is clearly stating that science in and of itself has no boundaries. It is free to follow the data wherever it may lead. He clearly states the boundaries on science in regard to materialism are not natural but have been placed there by people who subscribe to materialist ideology. This is another case of the struggle being ideological not scientific.

Biologist Scott Todd gives us another example of this when he makes this general statement,

"Even if all the data point to an intelligent designer, such a hypothesis is excluded from science."(10)

Why is it excluded? Whatever happened to 'following the data' regardless of where it leads? Why should the concept of intelligent design be automatically separated from science? When did this happen? Who made this distinction? Upon what grounds is it made? Clearly this is an ideological discussion and not a scientific one.

The idea that science can only deal with material things, atoms and molecules, is a relatively new phenomenon.

Historically, modern science grew out of Christian theology based on reason. Modern science was founded by Christians and nourished by them. Evidence of this is all around us. Virtually all of the great universities in the western world were founded by Christians: Harvard, Princeton, Yale, Oxford, Cambridge, Columbia, Brown, Dartmouth, Notre Dame, Loyola, Boston College, and the Universities of Paris, Montpelier, and Salamanca, just to name a few. Nearly all of the early pioneers in science: Copernicus, Galileo, Newton, Dalton, Faraday, Lavoisier, Kelvin, Pasteur, Mendel, Faraday, Maxwell and many more were Christians who believed they were unlocking the mechanisms of God's handiwork. They established science and the scientific method as we know and use it today.

Sir Isaac Newton speaks for them all when he states,

"This most beautiful system of sun, planets and comets could only proceed from the counsel and dominion of an intelligent and powerful being,"(11)

Nickolas Copernicus agrees when he said,

"So vast without any question, is the divine handiwork of the Almighty Creator."(12)

Christianity was able to lay the foundation for modern science because it is based on reason.

Thomas Aquinas, a Christian Friar, Priest, Philosopher and Theologian wrote in his Summa Contra Gentiles,

"We shall first try to manifest the truth that faith professes and reason investigates, setting forth demonstrative and probable arguments, so that the truth may be confirmed and the adversary convinced."(13)

St. Augustine gave us an example of Christian reasoning as he tackled the problem of time.

He reasoned, if before today there was yesterday and before yesterday the day before yesterday then, do the yesterdays extend infinitely into the past? If so, how could God have created the universe if it was infinite? If God created the universe it must have had a beginning. If God created the universe, what was He doing before He created it?

St. Augustine's answer was, God created *time* along with creating the universe and God exists outside of time. Today scientists understand the universe had a beginning and time began at that beginning. St. Augustine is right.

This type of reasoning was not previously found in Judaism, Islam, Hinduism, or Buddhism. Today scientists from all areas of life practice this type of reasoning.

We should remember, the beginning of modern science has its roots in Christianity and virtually all of the early modern scientists were strong believers in the Christian God. This is the history of modern science and Christians can be proud of it.

So, to what specifically is Weinberg referring when he says science is corrosive to religion? Is it Chemistry, Physics, Meteorology, Medicine? No. He refers to the indoctrination of students with Darwinian evolution. The idea that life is not created but is the product of blind random unguided forces. This is the idea that is corrosive to religion.

Weinberg and atheists in general would also have us believe the methodology of science is corrosive to religion, a belief that truth can only be revealed by science. This is an extremely shallow and superficial approach to the universe and life in it. This approach makes an enormous assumption, the assumption that there is nothing more to the universe than random arrangements of atoms and molecules.

The following chapters highlight advances in science that increasingly point to the need for a designing hand in the creation of the universe and life in it. Science is establishing the necessity for a Creator.

There is an ideological struggle for the hearts and mind of our children and their eternal future. The struggle is between Darwinian materialism and belief in God. Parents need to be aware of this struggle. It is my hope this book will provide the tools needed to resist the temptation to embrace the ideology of Darwinian materialism.

We are not alone in this struggle. We have some awesome scientific minds on our side.

The great French biologist Louis Pasteur said,

"The more I study nature the more I am amazed at the Creator" (14)

Francis Bacon, one of the founders of scientific inquiry, put it this way,

"A little science estranges a man from God, a little more brings him back."(15)

Albert Einstein intuitively knew what we know in our hearts and he beautifully expressed it in the following quote.

"I'm not an atheist, and I don't think I can call myself a pantheist. We are in the position of a little child entering a huge library filled with books in many languages. The child knows someone must have written those books. It does not know how. It does not understand the languages in which they are written. The child dimly suspects a mysterious order in the arrangement of the books but doesn't know what it is. That, it seems to me, is the attitude of even the most intelligent human being toward God. We see the universe marvelously arranged and obeying certain laws but only dimly understand these laws. Our limited minds grasp the mysterious force that moves the constellations."(16)

The following chapters will show how truthful and prophetic these statements are.

Before we continue it would be wise to review what atheistic Darwinian materialists believe:

- The universe and everything in it can *only* be explained by purely natural causes.

- There can be *no* God of any kind.

- There is *no* soul, *no* life after death, *no* free will, *no* absolute moral authority and *no* foundation for ethical behavior.

- The universe appeared out of nothing.

- Life *spontaneously* appeared from non-life.

Atheist Quentin Smith sums up Darwinian materialism nicely,

"The most reasonable belief is that we came from nothing by nothing for nothing."(17)

The struggle for the hearts and minds of our children is very real and very much alive. The main battleground of this ideological struggle is High School and University Biology classrooms where Darwinian evolution is taught.

It would be a serious misconception to categorize all Biology teachers as knowing and willing accomplices in this struggle. Most Biology teachers teach what the text dictates. Certainly some teachers may use evolution as a tool (I did) to discredit the Bible either directly or indirectly. Others just teach the text and are uninterested in the consequences. Still others may be aware of the underlying message and may even try to soften it. The message, however subtle, always imprints the student mind. The message is: you are not God created; you are a product of an undirected series of blind random mutations and natural selection.

How many thousands of High School Biology teachers go to church on Sunday then, on Monday morning, teach in their science class that life is a purely natural phenomenon? There is a serious disconnect here that only a few people and organizations choose to recognize or address. Atheists have convinced many of us that science and religion cannot mix. This is simply not true. Science can be studied from the perspective of unlocking Gods handiwork just as effectively as from the perspective of purely material phenomenon.

Many scientists have locked themselves into a materialistic box. Choosing not to think out of the materialistic box is a limiting factor for them. How can they be so sure there is nothing more to existence than atoms and molecules? Pure science follows the data/evidence wherever it leads. Failure to do this reminds us the differences are ideological not scientific.

There is no need to check our religious views at the door of science. It is atheistic propaganda that some scientists, teachers and parents have swallowed hook line and sinker. Atheists are ecstatic when this happens. It gives them a free hand in the instruction of our children.

When something is unknown in science, atheists promote the idea that Christians throw up their hands and say God did it. This is nonsense.

The Christian scientists mentioned earlier in the chapter never did that and Christian scientists today do not do it. A Christian scientist can make new and important discoveries of God's wonders just as efficiently and effectively as any other scientist, maybe even better!

Dr. Henry 'Fritz' Schaefer is the Graham Perdue Professor of Chemistry and director of the Center for Computational Quantum Chemistry at the University of Georgia. He understands the relationship between science and Creation when he states,

"The significance and joy in my science comes in those occasional moments of discovering something new and saying to myself, 'So that's how God did it.' My goal is to understand a little corner of God's plan."(18)"

It is in the area of Darwinian evolution that Christian and atheistic scientists differ. In reality it is the atheists who have chosen to block their ears, shut their eyes and close their minds to the possibility there could be more to existence than atoms and molecules.

Albert Einstein acknowledges the existence of more than atoms and molecules when he states,

"My religion consists of a humble admiration of the illimitable superior spirit who reveals himself in the slight details we are able to perceive with our frail and feeble minds. That deeply emotional conviction of the presence of a superior reasoning power, which is revealed in the incomprehensible universe, forms my idea of God."(19)

The Christian worldview, God as Creator, and Darwinian materialism are incompatible. Our children are taught in church that God is the Creator, they then go to school and are taught life arose spontaneously and thereafter developed by blind random mutation and natural selection. No wonder they are confused! There is no way both of these ideas can be true.

Our children are taught this over and over in grade school, high school and university. This confusion grows and many of them begin to question their beliefs and embrace Darwinian materialism. This has been going on for so long that parents, having come through the same indoctrination, are also confused and often think they can believe in both concepts at the same time.

Virtually every opinion poll taken in the U.S. confirms that 80 to 90% of Americans believe in a God. How did we get into this surreal situation where the overwhelming majority of Americans believe in a Creator yet, we allow our children to be taught life is a result of a random, natural, material process devoid of the supernatural?

This confusion leads many of us to believe that God may have put the wheels of Darwinian evolution in motion, then stepped back and let it roll in an unguided random fashion. This is the compromise all atheistic materialists want to see in us. It is a compromise that cannot be justified. It is trying to have the cake and eat it too. If you feel comfortable with this compromise here are two points to be aware of:

1. Darwinian materialism does not allow for any supernatural interference. If you believe God has any role at all in evolution, whether the starting of it or the arrangement of the universe to enable it, you do not believe in Darwinian materialism, you believe in intelligent design. If this is your situation, are you content to let your children be taught that life is purely a blind random natural occurrence?

2. If you pray, you must believe God hears prayer and prayer can make a difference. If God hears prayer and prayer can make a difference then God is active in the world. If He is active in the world then it is unreasonable to think He has no direct input in the history of perhaps his greatest achievement, the creation of life. What kind of a God numbers the hairs on our head and yet plays no part in the magnificent panorama of life.

There are a few well known Christian scientists who tread this thin line. They would have us believe God initiated the beginning then let the laws of nature take over and in time human beings were produced. If they are truly Christians, I assume they go to church. Churches are houses of prayer and sooner or later, people pray. The simple act of prayer acknowledges an active and participant God. If God is not active in life then what is the point of prayer? It is not possible to reconcile prayer and a non-participating God. Believing God initiated the beginning and then stood back to let nature take over is an unnecessary compromise. The following chapters will take a closer look at Darwinian evolution and will clarify why this compromise is unnecessary.

Chapter Three

UNDERSTANDING EVOLUTION

"Darwin's general theory of evolution may, in the final analysis, be little more than an unwarranted extrapolation from **micro**evolution based more on philosophy than fact."

Professor Art Battson (20)

Do you believe in evolution?

My answer is a resounding YES!

A second time.....

Do you believe in evolution?

My answer is a resounding NO!

Elaborating on these responses will explain the basic confusion surrounding evolution.

We often hear that evolution is a scientific fact. We read that virtually all scientists believe in evolution. These are powerful statements. Can they be true?

The answer is yes and no, because there are two of types of evolution.

MICROevolution and **MACRO**evolution

We can embrace **microevolution**, the 'yes' evolution, and have serious doubts concerning **macroevolution**, the 'no' evolution. It is scientifically valid to say yes to the one and no to the other. We are going to learn **microevolution** is scientific fact whereas **macroevolution** is not scientific fact. **Macroevolution** is often called a theory however it is a theory being slowly disassembled by modern science. **Macroevolution** is better classified as a hypothesis rather than a theory when viewed in light of modern scientific advances.

MICROEVOLUTION

Microevolution is confined to changes over time within a 'type' or 'kind' of creature. The words type or kind refer to groups of living organisms that have descended from the same ancestral gene pool. This means the mechanisms of **microevolution** have produced an increasing variety of clams, birds, fish, and even humans. It is important to note that although the clams change over time, they are still clams, birds are still birds, and so on. One type or kind does not change into another type or kind. The changes are only **within** a type or kind.

For example the fossil record shows the sudden appearance of phylum Annelida, the worms. Over time this group of creatures show a variety of changes however they remain worms throughout the fossil record. The same applies to all other types or kinds of creatures. These creatures appear suddenly, change and develop over time and yet like the worms they remain true to their original type. These changes within a kind are the only type of verifiable changes found in the fossil record. The fossil record will be examined in chapter 7.

Darwin's finch studies on the Galapagos Islands are well known. Over time the finches developed, amongst other things, different feeding habits. Some developed large beaks to crack larger harder nuts while others developed smaller beaks to feed on softer seeds. Although they changed over time, they began as finches and are still finches. This is a classic example of changes within a kind, **micro**evolution.

MACROEVOLUTION

Macroevolution theory assumes life began in some organic slime pool by organizing itself into molecules. The molecules became self- replicating and over time produced all life as we know it. This is the idea of descent with modification. **Macro**evolution is said to be change from one 'kind' to another. **Macro**evolution is the 'molecule to man' concept.

The following table shows a comparison between **micro**evolution and **macro**evolution.

Microevolution	Macroevolution
Changes **within a 'kind'** For example: Varieties of dogs or clams	Change **from one 'kind' to another** For example: Reptiles to Mammals
Results in a **loss** of genetic information	Requires a massive **gain** of genetic information
Is observable	Has never been observed
Is reproducible	Is not reproducible
Is accepted by all scientists	Is not accepted by all scientists

Microevolution (changes within a kind).

In Genesis 1: 11, 12, 21, 24, & 25, the Bible clearly states God created 'each **according to its kind.**'

The genetic makeup of all creatures contains a tremendous amount of variation. Just for a moment envision the many variations on the human form. However, despite these many variations, all are still human. Variation exists among all creatures. The important point to note is, for example, although there are many variations of dogs, they are all still dogs. These variations have come about by adaptations to different geological areas, climates or chemical environments. These variations are examples of **micro**evolution.

These variations can be readily reproduced and observed in the laboratory. Experiments on fruit flies and bacteria routinely produce only variations of fruit flies and bacteria.

Fifty thousand generations of breeding bacteria in a laboratory has produced only bacteria.

Virtually all scientists agree **micro**evolution produces change within a kind. This would include the breeding of animals like dogs, horses and pigeons. It is important to also note that try as we might there appears to be a limit to the amount of variation breeding can achieve. At some point the new variant becomes so inbred it cannot survive. **Micro**evolution has scientifically proven variation limits.

One final very important point. **Micro**evolution results in altered or lost genetic material. **No new genetic information** is produced. This is a very important point to remember when we examine **macro**evolution.

Microevolution is evolution we can believe. It is in line with God's word in Genesis, He says He created each according to its kind.

Microevolution is a scientific fact. When people say evolution is a scientific fact because bacteria or viruses mutate, you can agree with them. These are classic cases of **micro**evolution, changes within a 'kind.'

To sum up: **micro**evolution is restricted to changes within a kind.

Macroevolution requires changes **from one kind to another.**

As stated above, **macro**evolution theory assumes life began spontaneously in some organic slime pool. It then assumes life organized into molecules which became self-replicating molecules. The theory then assumes this original life form diversified into all forms of life that exist and ever existed. The array of assumed diversification is astounding! It ranges from bacteria to clams to spiders to birds to dinosaurs to humans.

Note that in **macro**evolution we see a change from one kind of creature to another. It is important to recognize that these changes require **huge** amounts of **new, never before existing,** genetic material.

Mammals for example, have a great deal of genetic information reptiles do not. The following table illustrates just some of the characteristics requiring genetic material mammals have that reptiles do not.

REPTILES	MAMMALS
Cold blooded	Warm blooded
Scales	Hair or fur
Egg laying	Born alive
No mammary glands	Mammary glands
No diaphragm	Diaphragm
Teeth replaced often	Teeth replaced once
Sex type by temperature	Sex type by chromosome

Macroevolution theory repeatedly requires huge gains in genetic material.

Macroevolution is a historical science constructed from assumed events and occurrences in the past. Therefore it is not observable or reproducible and as such, will always be theoretical.

Some scientists push the idea that descent with modification (**macro**evolution) is a scientific fact. It is not. It is at best, a theory.

Macroevolution is one theory of the origin and development of life. However on close inspection it is found to be a very shaky theory and definitely not a scientific fact.

Therefore one objective of this book is to discredit and disprove the idea that **macro**evolution can be classed as a scientific fact. This book will provide many examples and situations showing **macro**evolution cannot be categorized as scientific fact. It is better classified as a shaky theory. Chapter 10 will give a scientific alternative to **macro**evolution.

Proponents of **macro**evolution often use a technique commonly known as 'bait and switch.'

Here is a description of the technique known as 'bait and switch.' Perhaps you have had some experience with this type of marketing in the past.

Imagine an ad on TV for a computer at a really great price, lower than you have ever seen. You go to the store to buy it and find out that it is made by a company you never heard of. Then you find out that it has limited memory and low resolution. Immediately the salesperson shows you a different computer, much better and much more costly. You have been a victim of bait and switch.

You were never meant to buy the cheap computer. The advertisement got you into the store and the salesperson attempts to sell you what they intended to sell you, a much more costly computer. The cheap computer was the bait, you needed to be switched to the more expensive one.

All too often **macro**evolution is justified by examples of **micro**evolution. This form of bait and switch is especially prevalent in Biology text books. A strong case for **micro**evolution is made then the conclusion is drawn that we are products of decent with modification, **macro**evolution. Do not buy into this line of reasoning. A careful look at the table comparing and contrasting **micro**evolution and **macro**evolution shows the stark differences between the two.

The National Association of Biology Teachers (**NABT**) uses a subtle form of bait and switch in their official statement on the teaching of evolution.

I have reproduced their statement below and have highlighted a number of sentences. Under each paragraph I have commented on the highlighted sentences.

NABT Position Statement on Teaching Evolution is as follows:

"The frequently-quoted declaration of Theodosius Dobzhansky (1973) that 'Nothing in biology makes sense except in the light of evolution' accurately reflects the central, unifying role of evolution in the science of biology. As such, **evolution provides the scientific framework that explains both the history of life and the continuing change in the populations of organisms** in response to environmental challenges and other factors. Scientists who have carefully evaluated the evidence overwhelmingly support the conclusion that both the principle of evolution itself and its mechanisms best explain what has caused the variety of organisms alive now and in the past."

Author's comment:

Let's examine a section of the highlighted sentence above, '**evolution provides the scientific framework that explains both the history of life**.' This is most certainly referring to the **macro**evolution idea, change from one kind to another. **Macro**evolution does not 'explain' the history of life. It is but one hypothetical idea put forward to account for the history of life. Yet, when studied in detail under the light of modern technological advances, it becomes a shaky hypothetical idea. To say it 'explains' the history of life is nothing more than a huge assumption.

The last part of the highlighted sentence states '**evolution … explains … the continuing change in the populations of organisms.**' This is no longer an assumption, but rather solid **micro**evolution science, 'change within a kind.' This is the 'bait and switch.' The **micro**evolution idea of change in populations is inappropriately tied to the **macro**evolution idea of explaining the history of life. What we see in life is continuous change 'within a kind.' We do not see change from one kind to another.

Change occurring from one kind to another is a purely theoretical idea. Tying the two ideas, the history of life and continuing change in populations together, is using **micro**evolution to justify **macro**evolution. This is an example of 'bait and switch.'

The second paragraph in the **NABT** statement continues below.

"The principle of biological evolution states that **all living things have arisen from common ancestors.** Some lineages diverge while others

go extinct as a result of natural selection, mutation, genetic drift and other well-studied mechanisms. The patterns of similarity and diversity
in extant and fossil organisms, combined with
evidence and explanations provided by molecular biology, developmental biology, systematics,
and geology provide extensive examples of and
powerful support for evolution. Even as biologists
continue to study and consider evolution, they
agree that all living things share **common ancestors** and that the process of evolutionary change
through time is driven by natural mechanisms."

Author's comment:

The highlighted line, **all living things have arisen from
common ancestors**, is interesting because ancestors is used
in the plural. **Micro**evolution would agree that all living
things had common ancestors, 'each according to its kind.'
Macroevolution requires the statement to read in the sin-
gular, from a common ancestor. Later on in the same paragraph the plural, ancestors, is used again and thus the whole
paragraph is in accord with **micro**evolution, not **macro**-
evolution. If the authors of the **NABT** statement intend this
paragraph to reinforce the idea of **macro**evolution they are
doing so using **micro**evolution, another example of 'bait
and switch.'

On to the third paragraph of the **NABT** statement on
evolution....

"Evolutionary biology rests on the same scientific methodologies the rest of science uses,

appealing only to natural events and processes to describe and explain phenomena in the natural world. Science teachers must reject calls to account for the diversity of life or describe the mechanisms of evolution by invoking non-naturalistic or supernatural notions, whether called "creation science," "scientific creationism," "intelligent design theory," or similar designations. Ideas such as these are **outside the scope of science** and should not be presented as part of the science curriculum. **These notions do not adhere to the shared scientific standards of evidence gathering and interpretation."**

Author's comment:

The above paragraph places limits on the most important of all scientific principles, the principle that one must follow the data regardless where it leads. For example, data strongly suggests the universe is a product of design. Is the data to be ignored because someone arbitrarily decided it is outside the scope of science? Whatever happened to, 'follow the data?' Again, the data strongly suggests the origin of life on earth is best explained as a product of design. Should we explore the data or not?

Statements like 'outside the scope of science' are arbitrary and short sighted. They are a product of ideology rather than science. An ideology designed to ensure our children are taught a materialistic view of the origin of the universe and life. What we see in the above **NABT** paragraph is an atheistic bias masquerading as scientific method.

The above **NABT** paragraph could be summarized:

Follow the data wherever it goes, unless the data suggests there is something more to life than atoms and molecules. If the data suggests there is something more to life than atoms and molecules, stop, the data cannot be further considered.

The last highlighted sentence in the third **NABT** paragraph implies that intelligent design theory does not follow **shared scientific standards of evidence gathering and interpretation**. This is simply not true. I challenge the National Association of Biology Teachers to give an example showing the Discovery Institute of Intelligent Design not following solid scientific methods.

I doubt they have done much homework on this topic. They are simply promoting the atheistic bias in science.

Let's look at the fourth **NABT** statement paragraph:

"Just as nothing in biology makes sense except in the light of evolution, nothing in biology education makes sense without reference to and thorough coverage of the principle and mechanisms provided by the science of evolution. Therefore, teaching biology in an effective, detailed, and scientifically and **pedagogically honest manner** requires that evolution be a major theme throughout the life science curriculum both in classroom discussions and in laboratory investigations."

Author's comment:

The first reading of this paragraph seems to make sense. However there is an implied mindset that the mechanisms of evolution should not be tampered with or looked at in a critical manner. This is the very antithesis of scientific principles which dictate that anything and everything is open to discussion and critical analysis.

As for teaching evolution in a **'pedagogically honest manner,'** I would ask the reader to keep this phrase in mind and decide for yourself whether biology textbooks actually teach evolution in a **'pedagogically honest manner.'** There will be further reference to this phrase.

Let's take a look at the fifth and final **NABT** paragraph

"Biological evolution must be presented in the same way that it is understood within the scientific community: as a well-accepted principle that provides the foundation to understanding the natural world. Evolution should not be misrepresented as 'controversial,' or **in need of 'critical analysis'** or special attention for any supposed **'strength or weakness'** any more than other scientific ideas are. Biology educators at all levels must work to encourage the development of and support for standards, curricula, textbooks, and other instructional frameworks that prominently include evolution and its mechanisms and that refrain from confusing non-scientific with scientific explanations in science instruction."

Author's comment:

If there was ever any uncertainty about the closed mind of the **NABT** position statement on evolution, the above paragraph conclusively dispels it. Here are the reasons why.

Macroevolution is a shaky theory that:

1. cannot explain the origin of life,

2. cannot explain the origin of amino acids,

3. cannot explain the formation of proteins,

4. cannot explain the formation of DNA,

5. cannot explain the information content of DNA,

6. cannot explain epigenetic information,

7. cannot give one solid example of the mechanism of blind random mutation and natural selection at work in a **macro**evolution setting, and

8. cannot produce the hundreds of transitional fossils required to validate **macro**evolution.

AND YET......

The **NABT** statement tells us the **macro**evolution theory is not in need of critical analysis. How ridiculous! This statement is nothing but atheistic propaganda. It would be laughable if it were not being foisted on our children. They deserve the truth.

The reason for not wanting any critical analysis of **macro**evolution is simple. If our children were allowed to see a critical analysis, they would see **macro**evolution for what it is. They would see a shaky theory heavily reliant on assumption, interpretation and extrapolation.

The authors of the NABT statement on evolution might want to follow Charles Darwin's advice when he wrote in the Origin of Species, "a fair result can be obtained only by fully stating and balancing the facts and arguments on both sides of each question."(21)

The use of **micro**evolution to validate **macro**evolution is a common practice in evolutionary Biology. The reason for this is quite simple, there is little to no solid scientific data available to prop up **macro**evolution. Whereas, there is abundant solid scientific data to justify **micro**evolution. This is because **micro**evolution is 'hard' science, it is observable and reproducible. **Macro**evolution is 'historical science,' and as such it consists of the following.

1. Assumption: to take for granted, to *suppose* something to be a fact.

2. Interpretation: to have or show ones understanding of the meaning of, or to give *one's own* conception of information.

3. Extrapolation: to arrive at a conclusion by *hypothesizing* from known facts or observations.

When a proponent of **macro**evolution investigates the natural world, it is with the assumption decent with modification has taken place. The investigator assumes **macro**evolution to be a scientific fact. All data is then interpreted with this assumption in mind. Any data that contradicts this assumption is re-interpreted to fall within the bounds of the original assumption. The investigator now extrapolates the data to reinforce the original assumption.

This is not real science. It is unfortunate that generations of science graduates have been thoroughly indoctrinated into thinking this way. My hope is a new generation of science graduates will set aside the **macro**evolution assumption and begin thinking independently.

Some readers may think the above paragraph an exaggeration of biological thought when dealing with **macro**evolution. There is a simple way to test the validity of the above paragraph.

Take a section of solid data that contradicts **macro**evolution theory and interpret it in a manner contrary to **macro**evolution theory. How will this interpretation be received by the scientific establishment? Although some effort will go into refuting the interpretation, far more energy will be expended in attacking the author of the contrary interpretation. No variance from **macro**evolution is allowed.

Data can be re-interpreted and extenuating circumstances can be given as to why the data is not in line with **macro**evolution. However, there can be no hint the data could be interpreted to undermine the core of **macro**evolution. Any attempt to undermine **macro**evolution will incur the wrath of the atheistic scientific community. **Macro**evolution is the centerpiece of atheism and an attack on it is taken as an attack on atheism. Sadly, scientific bullying is rampant in today's scientific community.

The word Biology comes from two Latin words, Bio meaning life and logia meaning the study of. Biology is the study of life. The works of microbiologist Dr. Michael Behe and Dr. Stephen Meyer clearly show they have made important contributions to the study of life. For example Dr. Behe's work on irreducible complexity illustrates the mechanism of **macro**evolution is not sufficient to explain systems like the bacterial flagellum and the blood cascading system. To date there has been no scientific refutation of any consequence in regard to Dr. Behe's work. Dr. Meyer's two books, *Signature in the Cell* and *Darwin's Doubt*, raise solid scientific questions about **macro**evolution's ability to account for the beginning of life, the information content in DNA and the Cambrian explosion of life.

Atheists within the scientific community have gone to extreme lengths to stifle Behe and Meyer. They have responded to Behe and Meyer's works in a disturbing and hostile fashion. Personal attacks make up the bulk of their criticisms. The level of pettiness is downright embarrassing. They have even gone to the extent of putting pressure on Amazon Books to remove Dr. Meyer's book from the science section and place it in the religious section.

You can find dozens of quotes from prominent scientists having serious doubts as to the mechanism of **macro**evolution. What is interesting is these quotes are from the 1960's, 70's and 80's. Today a scientist expressing these kind of doubts will be condemned by these same atheists. Scientists discussing design alternatives to **macro**evolution have met with outright slander and academic punishment.

There have been a number of practicing scientists demoted or fired for even discussing design science. This is nothing but scientific bullying. Those involved in this kind of censorship should be ashamed of themselves. This is a new low point for science as an undertaking. The mentality of 'think like we think or get out' is terribly detrimental to science. Science does not need science police. Science needs the freedom to explore any and all avenues.

Here are some recent cases of scientific bullying. It is beyond the scope of this book to go into detail in regards these cases, however one can easily do a search of the following names and explore the details.

- NASA team leader David Coppedge

- Ball State Physicist Eric Hedin

- Georgia Southern University lecturer Tom McMullen

- Amarillo College instructor Stanley Wilson

- California State University researcher Mark Armitage

Apparently many of today's universities love diversity of sexual relations and diversity of sexual orientation, but they are not interested in diversity of opinion.

Remember, the struggle is ideological not scientific.

Assumption, Interpretation and Extrapolation.

The theory of **macro**evolution is based on assumption, interpretation and extrapolation. Here are examples showing assumption, interpretation and extrapolation.

First example:

Parts of this example are excerpted from an online article titled, 'A Whale Fantasy from National Geographic.' This article was written by Harun Yahya. The full article can be seen at: http://en.harunyahya.net/a-whale-fantasy-from-national-geographic/.

It is well worth reading.

A skeleton found in Pakistan was named Pakicetus which means *whale from Pakistan*.

Below is a drawing of Pakicetus taken from the previously mentioned online article. Note the streamlined swimming position with the forearms paddling and the hind feet stretched out.

The next diagram from the online article shows the **actual** skeletal structure of Pakicetus.

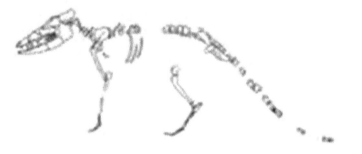

The skeleton shows no streamlined body, no indication of swimming, and feet designed for walking on land. These bones were found on land with bones of other land dwelling animals. They were found in conjunction with iron ore deposits, another sign of land dwelling animals.

Assumptions, interpretations and extrapolations about Pakicetus were mainly based on the structure of an ear bone and are simply not warranted.

The online article also shows a reconstruction, shown below, of the Pakicetus bones. This reconstruction gives a more realistic idea of the actual body of Pakicetus.

Pakicetus is best described as a four legged land mammal.

The diagram which showed Pakicetus with a streamlined body, swimming posture and broad hind feet is no more than a nice illustration of assumption, interpretation and extrapolation.

The same online article also shows a drawing of Ambulocetus, a creature supposedly close in line to Pakicetus in the assumed evolution of the whale. The name Ambulocetus means *walking whale*. These names are given to these obvious land animals in order to give some validity to the **macro**evolution idea that they are whale ancestors.

Note the streamlined swimming posture and the webbed feet.

Below is the actual skeletal structure of Ambulocetus. The legs are pressure bearing legs and are no more adapted to swimming than the legs of your pet dog. The webs between the toes are absolutely and purely imaginary.

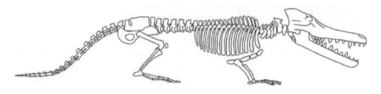

Pakicetus and Ambulocetus show every indication they were four legged land animals.

It is also interesting to note that in the above diagrams Pakicetus and Ambulocetus appear to be of about the same size. The fact is, Pakicetus is about the size of Ambulocetus's head. Ambulocetus is about ten times bigger than Pakicetus. Imagine a future ancestor of yourself being fifty or sixty feet tall. It takes a great deal of assumption, interpretation and extrapolation as well as imagination to accept that Pakicetus evolved into Ambulocetus.

The above descriptions of Pakicetus and Ambulocetus illustrate the blurring of science and wishful imagination. A classic example of assumption, interpretation and extrapolation.

A recent fossil find has muddied these waters even more. A skeleton very much like the skeleton of a modern whale was discovered and named the Antarctic Archaeocete. It was given an assumed age of 49 million years. The assumed age of Pakicetus is 50 million years and the assumed age of Ambulocetus is 48 million years. These assumed dates (a large assumption in itself) tell us three things.

1. It is quite likely Pakicetus, Ambulocetus and Archaeocete were alive at the same time. They were contemporaries not ancestors.

2. The assumed slow blind random mutation, natural selection process of **macro**evolution requires millions of years. Due to the very great differences in size and body characteristics, **macro**evolution theory must assume these three would have taken tens to hundreds of millions of years for one to evolve into the other. How could one have evolved into the other when they are dated as contemporaries?

3. These three creatures are so vastly different in size and shape there should be hundreds of transitional fossils between them in the fossil record. The transitional fossils are nonexistent.

Following is a list of characteristics requiring new genetic material needed to transform a land creature like Pakicetus into a fully seagoing whale. A tremendous amount of assumption, interpretation and extrapolation is needed to imagine all these essential changes happening in such a short time period or any time period. Here is a sample of the necessary modifications

New Genetic Material Needed:

- Ability to drink sea water (reorganization of kidney tissues)
- Forelimbs transformed into flippers
- Reorganization of skull bones and musculature
- Modification of the eye for underwater vision
- Counter-current heat exchanger for intra-abdominal testes
- Ball vertebra

- Tail flukes and musculature
- Blubber for temperature insulation
- Fetus in breech position (for labor underwater)
- Nurse young underwater (modified mammae)
- Reduction of hind limbs
- Reduction/loss of pelvis and sacral vertebrae
- Reorganization of the musculature for the reproductive organs
- Hydrodynamic properties of the skin
- Special lung surfactants
- Novel muscle systems for the blowhole
- Modification of the teeth
- Emergence and expansion of the mandibular fat pad with complex lipid distribution
- Modification of the ear bones
- Decoupling of esophagus and trachea
- Synthesis and metabolism of isovaleric acid (toxic to terrestrial mammals)
- Emergence of blowhole musculature and their neurological control.

The evolution of the whale as presented above is best described as a fairy tale.

The huge dependence **macro**evolution has on assumption, interpretation and extrapolation is further illustrated in the following example.

Second example:

Storrs L. Olson was the curator of birds at the National Museum of Natural History in the Smithsonian Institute.

When asked to comment on an article to go into the July 1998 edition of National Geographic titled, 'Dinosaurs take Wing.' Dr. Olson wrote the following letter clearly stating his views concerning the accuracy of the article National Geographic was going to print. Following is an excerpt from the letter Olson wrote to the head scientist at National Geographic.

"Prior to the publication of the article 'Dinosaurs Take Wing' in the July 1998 National Geographic, Lou Mazzatenta, the photographer for Sloan's article, invited me to the National Geographic Society to review his photographs of Chinese fossils and to comment on the slant being given to the story. At that time I tried to interject the fact that strongly supported alternative viewpoints existed to what National Geographic intended to present, but it eventually became clear to me that National Geographic was not interested in anything other than the prevailing dogma that birds evolved from dinosaurs.

Sloan's article takes the prejudice to an entirely new level and consists in large part of unverifiable or undocumented information that "makes" the news rather than reporting it. His bald statement that "we can now say that birds are theropods just as confidently as we can say humans are mammals" is not even suggested as reflecting the views of a particular scientist or group of scientists, so that it figures as little more editorial propagandizing. This melodramatic assertion had already been disproven by recent studies of embryology and comparative morphology, which, of course, are never mentioned.

More importantly, however, none of the structures illustrated in Sloan's article that are claimed to be feathers have actually been proven to be feathers. Saying that they are is little more than wishful thinking that has been presented as fact. The statement on page 103 that "hollow, hair-like structures characterize protofeathers" is non-sense considering that protofeathers exist only as a theoretical construct, so that the internal structure of one is even more hypothetical.

The hype about feathered dinosaurs in the exhibit currently on display at the National Geographic Society is even worse, and makes the spurious claim that that there is strong evidence that a wide variety of carnivorous dinosaurs had feathers. A model of the undisputed dinosaurs Deinonychus and illustrations of baby Tyrannosaurs are shown clad in feathers, all of which is simply imaginary and has no place outside of science fiction.

The idea of feathered dinosaurs and the theropod origin of birds is being actively promulgated by a cadre of zealous scientists acting in concert with certain editors at Nature and National

Geographic who themselves have become out-spoken and highly biased proselytizers of the faith. Truth and careful weighing of the scientific evidence have been among the first casualties in their program, which is now becoming one of the grander scientific hoaxes of our age- the pale-ontological equivalent of cold fusion. If Sloan's article is not the crescendo of this fantasia, it is difficult to imagine to what heights it can next be taken. But it is certain when the folly has run its course and has been fully exposed, National Geographic will unfortunately play a prominent but unenviable role in the book that summarizes the whole sorry episode.

Sincerely,
Storrs L. Olson."(22)

The complete letter can be seen at this site. https://answersingenesis.org/dinosaurs/feathers/sensationalistic-unsubstantiated-tabloid-journalism/

Dr. Olson's letter illustrates the magnitude of assumption, interpretation and extrapolation that pervades **mac-ro**evolution. In case anyone thinks the above exchange is merely a difference of opinion, keep in mind one of the creatures at issue was a dinosaur called Archaeoraptor. Shortly after the article was published, Archaeoraptor was found to be a 'scientific' hoax. National Geographic was forced to print a recantation.

It is impossible to read Dr. Olson's letter and not be struck by the amount of assumption, interpretation and extrapolation taking place.

The following example illustrates how assumption, interpretation and extrapolation can be used to take hard data and fit it into an evolutionary framework.

Third example:

> Professor J. C. Fentress of the University of Rochester observed that one species of vole (a mouse-like rodent) 'froze' when it observed a moving object overhead, while another species ran for cover.
>
> The species that froze in its tracks lived in the woodland, while the species which ran for cover lived in the open field.
>
> Professor Fentress told his colleagues (other Zoologists) about his observation, *but he purposely reversed the facts*, telling them that the woodland species ran for cover and that the meadow voles froze in their tracks.
>
> The other zoologists were able to give very elaborate and 'satisfactory' explanations why the woodland species ran and the meadow species froze, based upon conventional ideas of evolutionary theory.(23)

The fact that the zoologists gave elaborate and satisfactory explanations even though the facts had been reversed, is a clear example of assumption, interpretation and extrapolation.

Jerry Coyne, a Darwinian materialist, tries to convince us **micro**evolution absolutely leads to **macro**evolution. The very fact he tries to justify a link between the two indicates there is no scientific proof of **macro**evolution. There would be no need to try and link the two together if ample evidence of **macro**evolution were available.

This is the example he uses to justify linking **micro**evolution into **macro**evolution:

"When, after a Christmas visit, we watch grandma leave on the train to Miami, we *assume* that the rest of her journey will be an *extrapolation* of that first quarter-mile. A creationist unwilling to *extrapolate* from **micro**- to **macro**evolution is as irrational as an observer who assumes that, after grandma's train disappears around the bend, it is seized by divine forces and instantly transported to Florida."(24) (Bold italics are mine.)

Notice the key words, assume and extrapolate. Dr. Coyne admits that to move from **micro**evolution to **macro**evolution one must assume and extrapolate. If **macro**evolution is a scientific fact why must we assume and extrapolate, why not present the hard facts.

The truth is there are no hard facts, therefore one must assume and extrapolate to move from **micro**evolution to **macro**evolution. Dr. Coyne tries to convince us that anyone who is unwilling to assume and extrapolate is irrational, so irrational they would believe grandma's train is somehow magically transported to Florida. One succinct and amusing comment on his analogy went like this.

'It is really hard to know if grandma will ever arrive at Miami when she is laying the track, randomly directed, one rail at a time, as she goes.'

Macroevolution absolutely requires that grandma have no idea where she is going as she randomly lays the track. It is highly unlikely she would end up in Florida. Expecting she would, requires a magical quality to grandma's track laying.

Dr. Coyne has quite unwittingly shown us it is necessary to assume and extrapolate in order to explain **macro**evolution. This is the 'bait and switch' technique discussed earlier on. Unfortunately it is very subtle and not easily recognized by students.

Let us revisit the definition of extrapolate just to be sure we understand Dr. Coyne.

Extrapolate: to arrive at a conclusion by *hypothesizing* from known facts or observations.

Dr. Coyne by his own example clearly starts with the known fact of **micro**evolution and then hypothesizes **macro**evolution.

Macroevolution is heavily reliant on assumption, interpretation and extrapolation. It cannot be classed as a scientific fact.

In summary, **micro**evolution or change within a kind is evolution we can believe in. It is consistent with scientific evidence. It also is in accord with the Bible where it states God created each according to its kind.

Macroevolution is historical science, heavily reliant on assumption, interpretation and extrapolation. It is up to each of us to decide for ourselves just how good the theory of **macro**evolution is.

The following chapters will take a close look at **macro**evolution. The goal is to see if **macro**evolution is a solid enough theory to be classified as a scientific fact or whether it is a theory struggling to keep its position in science.

Chapter Four

THE BEGINNING OF THE UNIVERSE

"The more I examine the universe and the details of its architecture, the more evidence I find that the universe in some sense must have known we were coming."   Physicist Freeman Dyson (25)

"The exquisite order displayed by our scientific understanding of the physical world calls for the divine."
  MIT Physics Professor Vera Kistiakowsky (26)

"Lift up your eyes on high, and behold who hath created these things…"   Bible, Isaiah 40:26

"Hast thou not known? Hast thou not heard, that the everlasting God, the LORD, the Creator of the ends of the earth…"   Bible, Isaiah 40:28

A theory known today as the BIG BANG theory began to receive some prominence in the 1940's. Fred Hoyle, an astronomer, disliked the theory. He found the idea that the universe had a **beginning** to be pseudoscience resembling arguments for a Creator. Years later, Sir Fred Hoyle, now the Plumian Professor of Astronomy and Experimental Philosophy at Cambridge University, had this to say:

> "I do not believe that any scientist who examined the evidence would fail to draw the inference that the laws of nuclear physics *have been deliberately designed* with regard to the consequences they produce inside stars."(27

Hoyle's contemporary, Owen Gingrich, head of the Smithsonian Astrophysical Observatory, agreeing with Hoyle, said this:

> "Fred Hoyle and I differ on lots of questions, but on this we agree: "A commonsense and satisfying interpretation of our world suggests the designing hand of super-intelligence."(28)

The Big Bang theory was gaining solid acceptance by the late 1980's.

John Maddox, editor of Nature magazine, one of the top science journals in the world, wrote an editorial titled "DOWN WITH THE BIG BANG THEORY." He wrote:

> "... apart from being philosophically unacceptable, the Big Bang is an over-simple view of how the universe began and is unlikely to survive the decade ahead. Creationists seeking support for

their opinion have ample justification in the Big Bang."(29)

The Big Bang theory postulates a beginning to the universe. Maddox clearly understood, if the universe had a beginning it must have had a beginner. It is from this idea he states the Big Bang is philosophically unacceptable and Creationists can find support for their position.

Maddox dislikes the Big Bang theory and describes it as an over-simple view of how the universe began and hopes it will not survive future scientific scrutiny. Maddox wrote this article in 1989. He stated the Big Bang theory would not last the decade.

It is now nearly three decades later and although there are many points of contention about the Big Bang theory, almost everyone agrees the universe not only had a beginning but that it began in a flash of light. It began basically just as Genesis describes it.

"In the beginning God created the heavens and the earth." Genesis 1:1

The universe had a beginning.

"Then God said let there be light." Genesis 1:3

The universe began in a flash of light.

This is a clear case of the scientific data pointing to a Designer/Creator. Yet as far as Maddox was concerned if the scientific data points to a Designer/Creator it is 'philosophically unacceptable.' This is an ideological position, not a scientific one.

Robert Jastrow, founding director of NASA's Goddard Institute for Space Studies was a cosmologist and agnostic. A beginning to the universe had the following affect on him, he stated:

> "For the scientist who has lived by his faith in the power of reason, the story ends like a bad dream. He has scaled the mountains of ignorance; he is about to conquer the highest peak; as he pulls himself over the final rock, he is greeted by a band of *theologians* who have been sitting there for centuries."(30)

Jastrow refers to the opening lines of Genesis which clearly state the universe began in a flash of light, precisely as modern science has determined. Atheists still dislike the Big Bang theory because they understand anything that had a beginning had a beginner.

Atheists understand the scientific data. They just do not like the implications of the data. This is another example showing it is an ideological struggle not a scientific one. There is disagreement on various parts of the Big Bang theory but, almost universal agreement of a beginning in a flash of light just as Genesis describes it.

In an interview with *Christianity Today,* Jastrow said:

> "Astronomers now find they have painted themselves into a corner because they have proven, by their own methods, that the world began

abruptly in an act of creation to which you can trace the seeds of every star, every planet, every living thing in this cosmos and on the earth. And they have found that all this happened as a product of forces they cannot hope to discover. That there are what I or anyone would call supernatural forces at work is now, I think, a scientifically proven fact."(31)

Someone very ill goes to a doctor. His experience and training allows him to better understand health problems. So it is with the beginning of the universe. If someone wishes to understand it they need to seek the advice of astrophysicists and cosmologists, they have experience and training in this field.

Paul Davies, a physicist and cosmologist at Arizona State University, states:

"It is hard to resist the impression that the present structure of the universe, apparently so sensitive to minor alterations in numbers, has been rather carefully thought out. The seemingly miraculous occurrence of these numerical values must remain the most compelling evidence for cosmic design."(32)

Edward Harrison, professor of physics and astronomy at the University of Massachusetts, (Amherst), states:

"The fine tuning of the universe provides prima facie evidence of deistic design."(33)

Alan Sandage, the winner of the Crawford Prize in Astronomy, had this to say:

"I find it quite improbable that such order came out of chaos. There has to be some organizing principle. God to me is a mystery but is the explanation for the miracle of existence, why there is something instead of nothing."(34)

John O'Keefe an astronomer at NASA, makes this comment:

"We are by astronomical standards a pampered, cosseted, cherished group of creatures. If the universe had not been made with most exacting precision we could never have come into existence. It is my view that the universe was made for man to live in."(35)

The above quotes are a small sample of many available. It is worth taking a few minutes to read similar quotes in Appendix B.

The common thread through these quotes is the profound effect the organization of the universe has had on those most able to understand it, the astrophysicists and cosmologists.

Atheist try to say the universe is just a chance occurrence or just one of billions of universes possible. Keep in mind, all the above experts are well aware of these other possibilities. In spite of this they have rejected these other possibilities in favor of a designed universe.

The multi-universe idea is not science. Science is bounded by our universe and its laws. Anything outside our universe is not science but is pseudoscience or science-fiction. Multi-universes can be discussed by anyone, however the idea cannot be scientifically studied. It is beyond science.

Dr. Luke A. Barnes, a Super Science Research Fellow at the University of Sydney in Australia has this to say about the multi-universe idea:

"Could a multiverse proposal ever be regarded as scientific? Foft 228 notes the similarity between undetectable universes and undetectable quarks, but the analogy is not a good one. The properties of quarks–mass, charge, spin, etc.–can be inferred from measurements. Quarks have a causal effect on particle accelerator measurements; if the quark model were wrong, we would know about it. In contrast, we cannot observe any of the properties of a multiverse {M, f(m), pi}, as they have no causal effect on our universe. We could be completely wrong about everything we believe about these other universes and no observation could correct us. The information is not here. The history of science has repeatedly taught us that experimental testing is not an optional extra. The hypothesis that a multiverse actually exists will always be untestable."(36)

Atheists seem to lose their objectivity discussing the origin of the universe and our planet. They are happy to say our universe is one of billions of universes. They would say we won the universe lottery. Yet, when it comes to our planet they are just as happy to do a complete turnabout and say there is nothing special about our planet.

They say there are millions of planets like earth in the universe. A special universe and an ordinary planet like millions of others. In spite of the best interpretation of the data, they choose to deny the beginning of the universe and the very special nature of our planet. They instead embrace pseudo-science fiction ideas for which there is no data.

What is it about the universe that makes most astrophysicists start believing the universe is a product of design?

There are two main reasons for their belief.

The first reason, the universe had a beginning. It is commonly accepted in science and philosophy that anything having a beginning has a beginner. John Maddox understood this when in an earlier quote he said the Big Bang theory was "philosophically unacceptable." He was not prepared to accept a universe with a beginning.

Famous atheist philosopher David Hume is quoted as saying,

"I have never asserted so absurd a proposition as that anything might arise without a cause."(37)

Scientists generally agree the universe had a beginning. If the universe had a beginning it had a beginner. Scientists may not be able determine who, or what the beginner is, but Genesis states three important things:

1. The universe had a beginning,
2. It began in a flash of light,
3. God was the author of it all.

The second reason for believing the universe is a product of design, is the extreme fine tuning of thirty-four constants. A constant is a number expressing a property, quantity, or relation that remains unchanged under specified conditions. These constants control the destiny of the universe. They were established in the first milliseconds of the Big Bang. These constants are unimaginably fine-tuned. That's right, tuned so finely we as humans cannot even comprehend the degree of fine tuning.

Two examples of these constants are:

1. Gravitational force constant.

    ° If larger: stars would be too hot and burn too rapidly and too unevenly for life chemistry.

    ° If smaller: stars would be too cool to ignite nuclear fusion, thus many of the elements needed for life chemistry would never form.

2. Strong nuclear force constant

    ° If larger: no hydrogen would form, atomic nuclei for most life essential elements would be unstable, thus no life chemistry.

    ° If smaller: no elements heavier than hydrogen would form, again no life chemistry.

Find a complete list of the thirty-four constants in Appendix C

These are only two of the thirty-four constants. All the constants are so finely tuned as to mathematically eliminate the chance factor. The best scientific inference is that they were fine-tuned by a designer. There is no way of scientifically proving God as the designer, but that is not the point.

The best interpretation of current scientific data indicates the universe is a product of design.

One of these constants is called the cosmological constant.

The odds of the cosmological constant value being what it is are about one in one hundred million, billion, billion, billion, billion, billion. To put it numerically it is 10^{47} or

1 in 100,000,000,000,000,000,000,000,000,000,000,0 00,000,000,000,000.

No human can truly comprehend these numbers. Here is an illustration expressing the expanse this number represents. Imagine North America covered with dimes **all the way up to the moon**. Repeat this process **a billion times on another billion** North Americas. Color one dime red and place it randomly within the dimes. Finally, ask a blind man to pick out the red dime. The odds of him picking it out are 10^{47}.

Remember, this is only one of the thirty-four constants. Factor in the other thirty-three and the odds they were set randomly becomes so astronomical it is absolutely impossible for their exact settings to have been a random occurrence.

The existence and fine tuning of these constants has convinced most astrophysicists and cosmologists that our universe is not a random occurrence, but is a product of design.

Earlier in a quote from atheist philosopher David Hume he says it is ***absurd*** to think anything might arise without a cause. Note the word ***absurd***.

Science author Bill Bryson in his book '*A Short History of Nearly Everything,*' speaking on the beginning of the universe, states:

"It seems **impossible** that you could get something from nothing, but the fact that once there was nothing and now there is a universe is evident proof that you can."(38) (Bold print is mine)

Note the word *impossible*.

Hume's 'absurd' and Bryson's 'impossible' lead to the following conclusion.

Those who believe the universe began by itself, out of nothing, are left believing in the *absurdly impossible*!

Scientists now understand that the cause of this creation event is non-physical and non-temporal. The cause is super-natural.

If the universe is designed, then all that follows is a product of design. Design guided in detail by the finely tuned constants.

Chapter Five

THE ORIGIN OF LIFE

"Many investigators feel uneasy stating in public that the origin of life is a mystery, even though behind closed doors they admit they are baffled." Physicist, Paul Davies (39)

Until the 1860's scientists thought life began as a result of spontaneous reactions when conditions were just right. Then in 1864 Christian Scientist Louis Pasteur designed an experiment proving life only comes from other life.

This became the Law of Biogenesis:

Life only arises from life.

This is a very important biological law because, if life only arises from other life, how then did the very first life start? The simple answer to this question is, science does not currently know how the very first life began.

Scientists have no solid answer for the beginning of life, therefore **macro**evolution cannot be a scientific fact.

Macroevolution theorizes life somehow began spontaneously. This is highly theoretical and cannot pass as scientific fact.

It is amazing our children are taught the law of biogenesis on one day, life only arises from life, and the next day are taught life originated by some spontaneous reaction. All scientific evidence indicates life only arises from other life. Proponents of **macro**evolution, defying all the evidence, want us to believe life arose spontaneously. This is another case of the hard evidence being set aside in deference to ideology.

A proponent of **macro**evolution might believe life began by some kind of spontaneous generation, however, this is in direct *conflict* with the Law of Biogenesis and at this time there are no generally accepted, observable, scientific facts or experiments supporting how this could have happened.

Science author Bill Bryson in reference to a 1953 experiment says,

"Despite half a century of further study, we are no nearer to synthesizing life today than we were in 1953, and much further away from thinking we can."(40)

Over sixty years have passed since that experiment. There are still no generally accepted, observable, scientific facts or experiments supporting the spontaneous generation of the first life form. Proponents of **macro**evolution may believe this could have happened randomly, but scientific evidence indicates it is not feasible.

There are some serious problems associated with a random occurrence of life.

The basic unit of life is protein. Proteins are chemical strings of amino acids. There are twenty different amino acids available to form proteins in life. Obviously, in order to synthesize proteins, there needs to be a supply of amino acids.

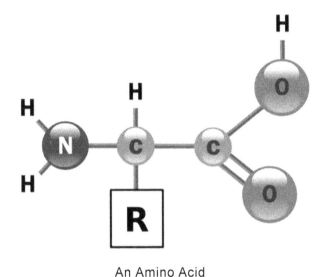

An Amino Acid

Now, let's examine the 1953 experiment mentioned by Bryson.

Two scientists named Miller and Urey undertook the 1953 experiment intending to show how amino acids came into being on the early earth. Scientists have known for many years it is a badly flawed experiment yet, it is still found in high school and college textbooks. Below is a diagram similar to the experimental apparatus they used.

A similar diagram can be found in many high school textbooks. This experimental apparatus usually appears under a heading of FIRST ORGANIC MOLECULES. These molecules are the amino acids needed for protein production.

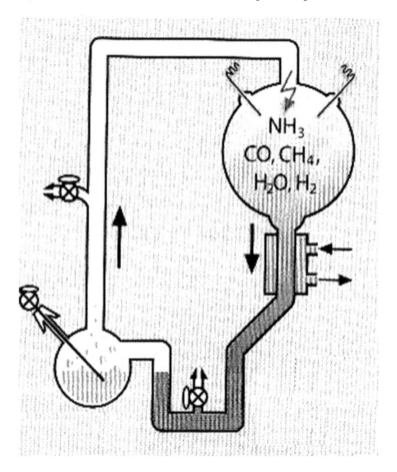

The average tenth grade student thinks the above experiment produces the organic molecules necessary to make proteins by a random process. The text does not tell students there are three serious flaws with the experiment.

The three serious flaws are:

1. The experimenters used the wrong gases.

Hydrogen, ammonia, methane and water were the gases used. Today, no one thinks there was any appreciable hydrogen in the earth's early atmosphere. Ammonia is very water soluble and quite unstable. It is unlikely ammonia was present in any appreciable amount in the early atmosphere. Without copious amounts of hydrogen and ammonia no amino acids could form. The experimenters assumed no oxygen was available. Recent studies indicate oxygen was available. Oxygen is very corrosive to amino acids and if amino acids formed, the oxygen would have degraded them.

2. The experimenters used the wrong methods.

The experimenters removed amino acids from the reaction chamber as they formed. The reaction then begins producing more amino acids because the original amino acids were removed. The experimenters are intervening in amino acid production. The production is no longer random.

Also, when the amino acids are removed from the reaction chamber this isolates them from other chemicals in the chamber. If the amino acids are left in the chamber they would cross react with other chemicals in the chamber and be destroyed. This is another case of the experimenters intervening in amino acid production.

If an early earth environment is being simulated, the amino acids cannot be removed to increase production and avoid degradation by cross reactions. Their intelligent intervention is necessary to insure amino acid formation.

The experimenters also intervened by cooling the apparatus and recirculating the gaseous mixture. The amount of amino acids formed would have been negligible without cooling and recirculating. It is highly unlikely in an early earth environment that cooling and recirculating would have been available. Few if any amino acids would have formed. Once again, their intelligent intervention is necessary to insure amino acid formation.

The net result of the experimenters tampering with the experiment shows that to produce and collect amino acids in this way it is necessary to utilize intelligent design. This experiment shows the need for intelligent intervention if production amino acids is the goal.

There have been a number of similar experiments undertaken since the Miller and Urey attempts, however they all suffer from the same problem, amino acids are only produced when the experimenters intervene (intelligent design.)

3. The experimenters got the wrong results.

Amino acids exist in two different forms, a right handed form and a left handed form. Each form can exist in four different shapes called alpha, beta, gamma and delta. For example, there could be an amino acid consisting of a left handed form with an alpha shape. There could also be a left handed form with a beta shape, or a gamma or delta shape. The same four types are possible using the right handed form. Altogether there are eight possibilities: Right-Alpha, R-Beta, R-Delta, R-Gamma; Left-Alpha, L-Beta, L-Delta, L-Gamma.

Proteins form when amino acids join together. There is no chemical reason for using only the left hand alpha shape or any other specific combination. All eight combinations are equal possibilities. Even though all possibilities are available, **life consists only of left handed alpha shaped amino acids.** There is no logical or chemical explanation for this. In fact one scientist, in frustration, commented 'God must have been left handed.'

The experiments discussed above isolated a combination of right and left handed amino acids in a variety of shapes. Since life only uses left hand alpha amino acids, any other type isolated would not be useful in forming proteins necessary for life.

Even these intelligent design type of experiments have only produced a limited number of amino acids and have never produced all the types needed to support life.

A more thorough explanation of problems associated with the above experiment is found on the following website. (http://www.truthinscience.org.uk/tis2/index.php/component/cobntent/article/51.html)

The Miller-Urey experiment is found in science textbooks under a heading of:

'ORIGIN OF ORGANIC MOLECULES'

The Miller /Urey experiment does not describe the *origin of organic molecules* by a natural, blind, random process. The heading is misleading if not downright deceptive. Yet, it is taught to our children as science. The truth is, scientists have no idea how amino acids could have randomly originated. **Macro**evolution cannot be a scientific fact if scientists have no valid idea how amino acids originated.

Recall the fourth paragraph of the National Association of Biology Teachers statement concerning evolution. It states biology should be taught in a "pedagogically honest manner." It is not honest to place the above experiment under the heading 'Origin of Organic Molecules' unless students are informed that not only were the wrong gases used but intelligent design was applied. Students will believe amino acids arose by random chance in an early earth environment unless they are specifically told that experimenters intervened.

The origin of amino acids is a chemical mystery. However, far more than amino acids are required to produce life. Amino acids, when correctly joined together, form proteins. Proteins are the building blocks of life. The origin of proteins is also a chemical mystery. The odds of proteins forming by random chance is illustrated by the following.

Imagine a large box containing hundreds of paper clips in twenty different colors. Each color indicates a different amino acid. Choose two hundred paper clips in a variety of colors. Join them together in a chain. Now, take the long chain and bunch it up into roughly a ball shape. You have made a protein model out of a paper clip chain. Of course in a real chain of amino acids the acids are not randomly joined together but rather in a very specific designated order.

The bunching of the chain also occurs in a very specific, designated order. It is the specific order of the amino acids and the specific order of the bunching that makes the construction of a protein virtually impossible by random process.

Every cell in every living thing contains protein. These proteins consist of left handed alpha shaped amino acids arranged as described above. Hemoglobin is a relatively short protein chain containing 146 amino acids. Collagen, a long chain, is made up of 1055 amino acids.

DNA (deoxyribonucleic acid) provides the blueprint for the production of proteins. Simply put, DNA directs production of proteins and proteins produce DNA. This is a classic chicken and egg scenario since without proteins you cannot make DNA and yet it is DNA that guides protein production.

Even if there were amino acids in an early earth scenario both the right and left handed forms of the amino acids would have been present. There is no chemical reason why both the right and left handed forms of each amino acid should not take part in reactions producing proteins. Yet life, in defiance of the chemical possibilities, uses only left handed alpha shaped amino acids to produce proteins.

These left handed alpha shaped amino acids are then arranged in very specific combinations depending upon which protein is being formed. For example, the odds of a two amino acid chain arranging itself in a specific order would be 20^2. This is 400 possible combinations since there are twenty different amino acids available. A four amino acid chain has 20^4 or 160,000 possible combinations.

The average length of a protein in living things is approximately 225 amino acids long. The possible combinations for a chain of this length is 20^{225}, an astronomical number. It is totally unreasonable to expect blind random chance to produce a protein of this length.

Bill Bryson in his book *A Short History of Nearly Everything*, comments on the odds.

"The chances of a 1055 sequence molecule like collagen assembling itself are, frankly, nil. It just isn't going to happen. To grasp what a long shot its existence is visualize a standard Los Vegas slot machine but broadened greatly to about ninety feet, to be precise- to accommodate 1055 spinning wheels instead of the usual three or four and with twenty symbols on each wheel (one for each common amino acid). How long would you have to pull the handle before all 1055 symbols came up in the right order? Effectively forever. Even if you reduced the number of spinning wheels to two hundred, which is actually a more typical number of amino acids for a protein, the odds against all two hundred coming up in pro-scribed sequence are 1 in 10^{260} That is a '1' followed by 260 zeros. That is a larger number than all the atoms in the universe."(41)

So, Bryson is convinced amino acids did not form into proteins by blind random chance. He then explains how Richard Dawkins in his book *The Blind Watchmaker,* proposes how amino acids may have assembled. Bryson relates that Dawkins believes there *must* have been *some kind* of cumulative selection process that allowed amino acids to assemble in chunks. *Must* have been *some kind of* cumulative selection does not sound like scientific fact to me. Dawkins' cumulative selection concept is a huge assumption. How can **macro**evolution be a scientific fact when there 'MUST have been SOME KIND of' cumulative selection? Do you see 'assumption, interpretation and extrapolation' in this phraseology?

The use of left handed alpha shaped amino acids and the specific arrangement of these amino acids to form proteins is a chemical mystery. At this time only the Creator knows the answer.

A dehydration reaction occurs when amino acids join to form proteins. This means in the process of joining amino acids together a water molecule is formed and released, leaving open bonding spaces. These open bonding spaces allow the amino acids to link up forming a chain. Thus, proteins could not have formed in a water environment like a warm slime pond, because dehydration reactions cannot take place in a water environment. The often heard story of life beginning in a warm slime pond is just not chemically feasible.

Many articles explaining the origin of life from a materialistic viewpoint suffer from the same problem. They grossly over-simplify the chemical processes required for life.

The authors of these 'origin of life' articles often show how a couple of organic molecules join together spontaneously. They then assume this could go on until a life form appears. However, these organic molecules come together following known chemical laws. The actual formation of amino acids into proteins does not follow any known chemical 'line-up' laws. It is an information driven process. This is analogous to comparing apples to oranges and is therefore an invalid comparison. Furthermore, authors often inject intelligent design into their origin of life scenarios. These articles rely on oversimplification and assumption, requiring a great deal of 'faith' to believe.

Finally, life consists of much more than a few chemicals joined together. For example, the simplest known form of free-living (non-host dependent) life, *Pelagibacter ubique,* is incredibly complex. It has *1,308,759* base pairs of DNA. This is the smallest number of DNA base pairs in a free-living life form and yet, it is over one million.

The high odds of forming proteins by random chance can be illustrated with a fun and simple game described in Appendix D.

The assembly of amino acids into proteins could not have taken place by blind random chance. The overwhelming odds against this force even committed atheist and Neo Darwinist, Richard Dawkins to suggest there MUST have been SOME KIND of cumulative selection process enabling amino acids to form proteins. Dawkins realized the assembly is not random but he is unable to explain how it happened. There is no life without proteins.

Francis Crick (Nobel Prize winner for discovering the structure of DNA) also realized the odds of life forming on earth by random chance are virtually impossible. His solution was to put forth the theory of 'pangenesis,' the idea that life was seeded from outer space. This begs the question; what is the origin of this 'outer-space' life?

Modern science has essentially eliminated the possibility of life forming spontaneously on earth. As a result, expect a variety of theories about how life was seeded from outer space.

A few meteorites have been found containing a couple of amino acids. No one is sure if these amino acids result from contamination or if they were actually present in the original meteor. The prospect of life occurring spontaneously on earth has become highly unlikely. Proponents of **macro**evolution now find it necessary to grasp at any straw to avoid the creation idea. It is a huge assumption to infer a meteorite containing a couple of amino acids suggests life was seeded from space.

There is no life without the highly complex proteins described earlier.

The random origin of amino acids, proteins and DNA appears to be chemically and statistically impossible. Proponents of **macro**evolution, realizing this, have proposed a compound called ribonucleic acid (RNA) as the precursor of life.

Dr. Stephen Meyer in his excellent book '*Signature in the Cell*' has devoted a chapter to RNA. In doing so he has shown RNA suffers from the same problems as DNA.

Since 'protein first,' 'DNA first' and 'RNA first' theories have all failed to explain the origin of life, **macro**evolution cannot be a scientific fact.

Professor of biochemistry Garret Vanderkooi gives a clear statement on current origin of life research;

"In the past, evolutionists were confident that the problem of the origin of life would be solved by the new science of Biochemistry. To their dismay, the converse has occurred. The more that is learned about the chemical structure and organization of living matter, the more difficult it becomes even to speculate on how it could have developed from lower forms by natural processes. From the scientific point of view, **evolution may have been a plausible hypothesis in Darwin's day, but it has now become untenable, as a result of fairly recent developments in molecular biology**" Biochemist, Garret Vanderkooi, (42)

Chapter Six

THE MECHANISM OF MACROEVOLUTION

"To postulate that the development and survival
of the fittest is entirely a consequence of chance
mutations seems to me a hypothesis based on
no evidence and irreconcilable with the facts.
These classical evolutionary theories are a gross
oversimplification of an immensely complex and
intricate mass of facts, and it amazes me that
they are swallowed so uncritically and readily,
and for such a long time, by so many scientists
without a murmur of protest." Nobel Prize winning Biochemist   Sir Ernst B. Chain (43)

Any theory must have a mechanism to justify it. A mechanism must show how the theory works. It must provide
a valid scientific pathway to achieve the end(s) the theory
is promoting.

Charles Darwin promoted a mechanism he thought
answered the question, 'why do living things appear to be

designed?' Be assured, most everyone in science agrees that living things appear to have been designed. Evidence of the Designer can be seen in the organization of the universe and in the immense complexity of living things. However, since designed entities require a designer, atheists dismiss the obvious. Darwin believed natural selection was the answer. He explained that living things appear to be designed and he believed natural selection was the designer.

The science of genetics was not available to Darwin. When Christian scientist, Gregor Mendel published his genetic studies it became widely accepted that Darwin was wrong. Darwinists seldom mention this little known fact. Today it is understood natural selection may play a part in **survival** of an organism but plays no part in its **arrival**. That said, there are no Darwinists left in biology.

Natural selection produces varieties of pigeons, variations of beak size in Galapagos' finches, changes in fox ears or rabbits' feet. These changes involve already existing genetic material. They are **micro**evolution changes. Natural selection only works on genetic material already present. To put it simply, there must be something to select before natural selection can occur.

Natural selection cannot produce the new genetic material required to bring about change from one kind to another. For example, the material needed to change reptiles to mammals. Natural selection is not capable of producing new genetic material.

Darwinists needed a mechanism to produce new genetic material. They now believe blind random mutations can create new genetic material. At this point they changed their name to Neo Darwinists.

The modern day proposed mechanism of **macro**evolution is one of blind random mutation and natural selection. Can blind random mutation and natural selection create the huge amount of new genetic material needed to change the assumed first replicating molecule into a human? This new mechanism needs careful examination.

Mutations are rare, estimates range from one in 10 million to one in 100 million duplications of DNA. Overwhelmingly, these mutations are either harmful or neutral. Unfortunately one mutation has very little effect. In order to create a new structure or system, dozens to hundreds of mutations all working toward the same end must take place. This is where the mathematical odds start to become prohibitive. Whereas, a single mutation has the odds of one in 10^7 (1:10,000,000) two related mutations have the odds of one in 10^{14} and three related mutations have the odds of 10^{21} and four related mutations have the odds of one in 10^{28}.

The odds of four related mutations taking place are extreme to the point of being statistically impossible. There is absolutely no reason a mutation would have anything at all to do with a previous mutation. This is because each mutation would be purely blind and random. Yet, four mutations are not nearly enough to begin to change a frog into a reptile. Some say, what about time, surely given enough time mutations could accumulate and make significant changes. Unfortunately, that is not how living systems deal with mutations. Unless each change (mutation) has some benefit, the organism attempts to eliminate it. Living systems do not store up mutations hoping that sometime in the future another one will occur and improve it.

Mutations corrupt, they do not create. Over thirty-five hundred mutational disorders have been identified in the human body. Studies on fruit flies over long periods and many generations have resulted in over three thousand mutational changes, all negative. These mutations are all **micro**evolution changes not **macro**evolution.

Dr. Lynn Margulis, famous biologist and author, clearly understood the relationship of random mutations and their connection to **micro**evolution, not **macro**evolution, when she stated,

"Although random mutations influenced the course of evolution, their influence was mainly by loss, alteration, and refinement... Never, however, did that one mutation make a wing, a fruit, a woody stem, or a claw appear. Mutations, in summary, tend to induce sickness, death, or deficiencies. No evidence in the vast literature of heredity changes shows unambiguous evidence that random mutation itself, even with geographical isolation of populations, leads to speciation."(44)

The few examples of mutations theoretically creating new genetic material are not convincing. These examples mainly describe tiny variations on an existing protein. Proponents of **macro**evolution believe these small theoretical variations could produce the huge amount of new genetic material needed to propel **macro**evolution. There is no solid scientific data to support this idea. It is nothing more than a huge assumption. This is the scientific equivalent of grasping at a straw.

When scrutinized, these small theoretical variations are found to be examples of modification or loss of function of an existing gene. This means an existing gene has been slightly altered or genetic material has been lost. In either case, no new, never before existing genetic material, has been produced.

It is important to remember these hypothetical small changes in a single protein are a long way from creating a new structure or system. The idea that these small theoretical changes can produce the vast amounts of new genetic material needed to change a reptile into a mammal is a purely hypothetical idea. It is certainly not a scientific fact.

Advantageous mutations confer some function which is useful and helpful in survival. For example, certain beetles on low windswept islands have lost their wings due to a mutation. This is considered an advantageous mutation. Flying beetles are often swept out to sea and end up drowning. A mutation causing beetles to lose their wings could confer survival advantages and could spread through the population. This would be an advantageous mutation. It would aid in survival. However, it is not an example of new genetic material. It is an example of loss of function. It is **micro**evolution not **macro**evolution. Whenever confronted with the term advantageous mutation, look deeper and you will find the advantageous mutation is a result of modification or loss of function; making it an example of **micro**evolution not **macro**evolution.

The genes that regulate body plan construction have been experimentally subjected to mutations. Invariably, these mutations destroy or severely damage the animal form as it develops from an embryonic state. How can mutations provide material for selection when they have been shown to destroy the organism? These experiments are not theoretical, they provide real hard data. Proponents of **macro**evolution believe mutations provide the material for selection. The actual experimental data tells us mutations do not, they destroy or severely damage the organism.

Whether lethal or damaged, the result is; there is no mutational variation for selection. If there is no selection there is no **macro**evolution. This conclusion is not supposition or assumption; it is based on solid experimental data.

Proponents of **macro**evolution assume **macro**evolution progresses by mutational changes in protein coding sequences. This means they assume a mutation acting on a protein will change the makeup of the protein. A long series of these changes might then result in a new organ or system and eventually a new body plan. The experimental work discussed above indicates this is not possible.

Mutation may affect genes controlling minor variations such as flower color or bacterial resistance to antibiotics. These are **micro**evolution changes not **macro**evolution changes. Mutations affecting major genes that theoretically could result in **macro**evolution change are not tolerated by the organism.

Richard Dawkins in his book 'The Blind Watchmaker' totally relies on mutation and what he calls cumulative selection to produce new body plans. The experimental work discussed above negates his theory.

Recent advances in developmental biology have shown that information guiding early embryonic development does not come solely from DNA. Information also comes from other parts of the cell. This information is crucial and is not related to DNA mutations. It is called epigenetic information, meaning information from outside the entity's genes. This means that generating a new body plan requires more than just genetic information. Even if mutations could successfully alter major genes, this alone would be insufficient to produce new body plans.

This recent experimental work provides two serious impediments to **macro**evolution.

1. Mutations in the critical embryonic form are lethal to the organism.

2. Embryonic development requires information from sources other than the organism's DNA. Epigenetic information from sources which are not susceptible to mutation.

The above experimental work raises serious doubts about mutation and natural selection's ability to produce the variety of life forms on earth. Once again, intelligent design becomes the best source of information needed to generate new body plans.

Complexity

Recent discoveries in microbiology have brought to light the immense complexity found in even the simplest of life forms. Bacteria are some of the earliest and simplest known life forms. Bacterial cells are incredibly complex. It is important to dwell on this complexity for a moment.

Bacterial cells are not just complex, they are complex beyond anything imaginable. They are astoundingly complex, beyond human engineering abilities.

For example, some bacteria have certain peptides that cooperate. These peptides act as *computational machines* performing *divisional operations* determining whether the food supply will sustain them sufficiently for successful reproduction.

Bacteria also use internal *proton pumps* to maintain a potential of 150 to 200 millivolts across their membranes. They are generating an electric current!

Bacteria also form *architecturally complex communities*. Cells migrate within these communities finding optimal reproduction sites.

Keep in mind bacteria are some of the simplest and earliest known life forms. These are just a few of many examples of complexity found in bacterial cells.

Notice the words researchers are using to describe these complex arrangements. Computational machines, architecturally complex communities and proton pumps are descriptions of designed systems, not blind random mutational ones.

Once again, it takes an incredible amount of assumption, interpretation and extrapolation to believe a process of blind random mutation and natural selection could produce computational machines, architecturally complex communities and proton pumps in the earliest known life forms. It is obvious living organisms look designed. The astounding complexity of bacterial cells confirms design in living things.

Microbiologist Dr. Michael Behe has identified a number of what he calls irreducibly complex systems in living organisms. These systems consist of a several parts that on their own have no value to the organism. They must all be there, working as a unit to be any benefit. They appear to be designed systems. There is no reason for an organism to keep any single part since each part on its own is of no value.

The bacterial flagellum is one example. The bacteria uses the flagellum (tail) for propulsion. The flagellum is driven by a microscopic electric motor. This motor uses basically the same design found in man-made electric motors. The bacterial motor has a rotor and a stator. It generates its own electricity, is self-repairing, and can go from forward to reverse faster than anything designed by man.

What good is the tail without the motor? What good is the rotor without the stator? Unless all these parts developed together the cell would have no reason to keep each of them separately. These parts are only useful as a unit. The odds of blind random mutation and natural selection producing all of these many parts simultaneously are statistically impossible.

Once again, we see the confirmation of design in living things. The confirmation of design is not based on supposition. It is the best inference from known scientific facts.

A few rather weak attempts have been made to discredit Behe's work on irreducible complex systems. None has any real substance. Yet, because the concept of irreducible complexity cannot be incorporated into **macro**evolution, it is basically being ignored. This is another case of ideology taking precedence over scientific data.

A multitude of systems within living creatures can only be described as machines. A machine is an intricate combination of interrelated parts most of which are absolutely essential to the machine's function. Remove an essential part and the machine breaks down. Remove or try to replace an essential part while the machine is running and chaos follows, likely destroying the machine. Yet this is exactly what a proponent of **macro**evolution would have us believe. They would have us believe mutations could alter a major part of a running organism without disrupting its function. This not only goes against common sense it contradicts what has been observed.

Cracks are appearing in the castle walls of **macro**evolution. The advances of microbiology, specifically technological advances, have allowed detailed observation into the innermost workings of living cells. The astounding complexities found in cells have motivated many microbiologists to look beyond blind random mutation and natural selection to discover how these complex systems could have developed. They are realizing blind random mutation could not possibly create these highly complex systems.

The noted microbiologist James Shapiro seems to believe the traditional blind random mutation, natural selection model cannot be the main source of evolutionary variation. He now talks about horizontal DNA transfer, interspecific hybridization, genome doubling and symbiogenesis as being the means of genetic variation. So, how do these different systems come together to create new genetic material?

Shapiro responds that natural genetic engineering can do it. Currently, no one is quite sure what natural genetic engineering is or how it works. However, natural genetic engineering implies the cell has the ability to generate the new systems it needs. Yet in order to generate the new systems, the cell must somehow know what it is going to need.

Where would the cell get this information? Blind random mutation and natural selection do not and cannot look ahead to see what the cell needs. An intelligent designer looks ahead to see what is needed and plans for that need.

Even assuming the cell somehow has the ability to look ahead to decide what new systems it needs, the structures it would then use to build these new systems are themselves immensely complicated.

Are we to believe the cell looks ahead, decides what new systems it needs, and then builds the structures required to assemble the new systems? If so, this scenario is vastly beyond the capabilities of blind random mutation and natural selection.

Once again intelligent design is more feasible.

Imagine the following scenario:

Imagine the first creatures with bones. One suffers a broken bone. The bone needs to heal. Healing occurs when new bone forms fusing the gap, repairing the break. This process requires unique instructions, chemical messengers and specific proteins. If these are not available, the bone will not be repaired and the chance of survival is minimal. What is the origin of these instructions, chemicals and proteins? They were not needed until the bone was broken. It is necessary to repeat the previous sentence. They were not needed until the bone was broken.

The creature cannot experience the bone breaking until it actually breaks. Once the bone breaks the creature will not survive unless repair is undertaken. The repair mechanism must be in place before any bone is broken. Only an intelligent designer can look ahead, foresee the need for bone repair and have the mechanism in place.

Mutation and natural selection could not have produced the instructions, chemicals and specific proteins necessary for repair unless mutation and natural selection somehow understood in advance that bones might be broken and would need repair. However, under no circumstances can mutation and natural selection plan ahead.

Macroevolution theory would clearly require millions of years for mutation and natural selection to produce the instructions, chemicals and proteins needed. Until a break occurs the creature would have no need for a repair mechanism and once the break does occur the repair mechanism must be immediately available. **Macro**evolution cannot explain the existence of the repair mechanism.

The repair mechanism appears in the creature's DNA before it is needed. Design is the most reasonable explanation for the existence of the repair mechanism.

New collateral coronary arteries and new scar tissue are other examples of systems needing advance instructions.

The immense specified complexity of living cells is causing a number of scientists to have serious doubts about the ability of blind random mutation and natural selection to produce complex genetic material.

If the basic mechanism of **macro**evolution cannot be demonstrated, **macro**evolution cannot be a scientific fact. And if, in the light of new advances in technology **macro**evolution is unable to account for the specified complexity found in living cells then, **macro**evolution cannot be a scientific fact.

Chapter Seven

THE FOSSIL RECORD

"Why is not every geological formation and every stratum full of intermediate links? Geology surely does not reveal any such finely graduated organic chain, and this, perhaps, is the most obvious and serious objection which can be urged against the theory. The explanation lies, as I believe, in the extreme imperfection of the geologic record."

Charles Darwin (45)

Contrary to what people think, the fossil record has *always* been the *weakest link* in the case for **macro-**evolution. Charles Darwin devoted a whole chapter to the fossil record in 'The Origin of Species.' His chapter titled, 'On the Imperfection of the Geologic Record' contained the above quote (45).

Darwin was well aware the fossil record did not agree with his implied idea of slow transition over time from molecule to man.

Blind random mutation and natural selection, the mechanism discussed in chapter 6, requires a fossil record of 'finely graduated organic chain.' The importance of this cannot be stressed enough. If, 'blind random mutation and natural selection' comprise the engine of **macro**evolution, their products should be clearly represented in the fossil record. They are not. The fossil record did not show in Darwin's time, nor does it show today, any 'finely graduated organic chain.'

Can blind random mutation and natural selection really create the tremendous amount of new genetic material needed to change a reptile into a mammal? Proponents of **macro**evolution would say this is oversimplifying the matter. They say reptiles did not change into mammals overnight. It was a slow gradual process with many transitional forms. If this is the case, where are these transitional forms? Are they in the fossil record? They are not. **Macro**evolution requires **hundreds** of transitional fossils. The multitude of transitional fossils required to verify change from one type to another exist only in the imagination of **macro**evolution proponents.

Remember, a single assumed transitional fossil is of absolutely no value when discussing finely graduated organic change. Hundreds of actual fossils are needed. Dragging out one or two assumed fossil examples merely illustrates the reality of the situation. The reality is; there is no fossil evidence of finely graduated change from one kind to another.

The eminent Harvard paleontologist, Dr. J. Gould, was well aware of this when he stated,

"At the higher level of evolutionary transition between morphological designs, gradualism has always been in trouble, though it remains the 'official' position of most western evolutionists. Smooth intermediates between Bauplane creatures are almost impossible to construct, even in thought experiments; there is certainly no evidence for them in the fossil record (curious mosaics like Archaeopteryx do not count)."(46)

The use of a few assumed transitional fossils to justify the need for hundreds is similar to the assumptions in chapters five and six. This is now the third time proponents of **macro**evolution have used huge assumptions to avoid the failure of **macro**evolution theory.

- Huge assumption number one: a few meteorite amino acids are used to suggest life was seeded from space.

- Huge assumption number two: a few contested instances of small mutations on a protein are used to justify the development of the complex machinery found in all living creatures.

- Huge assumption number three: a few contested specimens of transitional fossils are used to justify the hundreds of transitional fossils needed to validate **macro**evolution.

Macroevolution relies upon the acceptance of these assumptions. Thus, considerable audacity is necessary to claim that **macro**evolution is a scientific fact.

Although Darwin could see the obvious problem, no finely graduated change, he felt it was just a matter of time until the necessary fossils were discovered. Remember in the introductory quote Darwin said, 'the explanation lies, as I believe, in the extreme imperfection of the geologic record.'

This is not the case. Dr. Raup, a leading paleontologist and the curator of the Chicago museum of natural history, made this comment,

"Well, we are now about 120 years after Darwin, and knowledge of the fossil record has been greatly expanded, *ironically,* we have even fewer examples of evolutionary transition than we had in Darwin's time. By this I mean some of the classic cases of Darwinian change in the fossil record, such as the evolution of the horse in North America, have had to be discarded or modified as a result of more modern information."(47)

Darwin published 'The Origin of Species' in 1859. It is now 155 years after Darwin and 35 years after Dr. Raup's comment. Nothing has changed to alter Dr. Raup's opinion.

Dr. Raup is not alone, Harvard paleontologist Dr. J. Gould states,

"The extreme rarity of transitional forms in the fossil record persists as the trade secret of paleontology. The evolutionary trees that adorn our textbooks have data only at the tips and nodes of their branches; the rest is inference however reasonable, not the evidence of fossils."(48)

Dr. N. Eldredge, curator of the American museum of natural history gives this opinion,

"In the last decade, however, geologists have found rock layers of all divisions of the last 500 million years and no transitional forms were contained in them. If it is not the fossil record that is incomplete then it must be the theory."(49)

This quote from Dr. Eldredge was excerpted from an article he wrote which was titled, 'Missing, Believed Non-Existent.' How clear can it be!

Paleontologists, those who best understand the fossil record, generally agree the fossil record is basically devoid of any **macro**evolution transitional fossils.

Appendix E provides a list of quotes from paleontologists who agree that **macro**evolution transitional fossils are missing.

How do Biology textbooks deal with transitional fossils?

The following statement is similar to statements found in many biology textbooks.

Hundreds of transitional fossils have been discovered which document various intermediate stages in the evolution of modern species from organisms that are now extinct.

Are there hundreds of transitional fossils or not? The paleontologists quoted in appendix E have little faith in the existence of transitional fossils. Yet, biology texts tell children there are hundreds of transitional fossils. What are we to believe? Perhaps the better question is, what will our children believe when that is what they are taught?

If there are hundreds of transitional fossils, then the quotes from the many paleontologists would be wrong. Paleontologists are experts in the field of fossils. If an expert opinion concerning fossils is needed, paleontologists are the professionals to consult. It is unlikely the textbook authors are paleontologists. It is generally wise to go with the opinion of the experts.

Why would textbooks contain statements like the one above which is so contrary to the opinion of expert paleontologists? The answer is quite simple. Microevolution is again being used to justify **macro**evolution, another case of bait and switch.

Sure, there are hundreds of transitional fossils but they are all fossils supporting **micro**evolution (changes within a kind). They are **not** fossils supporting **macro**evolution (changes from one kind to another). For example, the fossils under discussion could be fossils of ancient squids, being compared to squids alive today. Some changes may be noted; however the comparison is between ancient squids and current squids. Another common example refers to Darwin's finches. Although they showed changes over time, they were finches when they came to the Galapagos Islands and they are finches today. These are comparisons within a kind, not from one kind to another. It is **micro**evolution not **macro**evolution. Textbooks often describe a number of such examples but in every case the example is **micro**evolution not **macro**evolution.

Textbook authors 'believe' **micro**evolution leads to **macro**evolution. They may 'believe' this but they do not document it, they rely on **micro**evolution examples.

Their 'belief' that **micro**evolution leads to **macro**evolution becomes obvious When textbook chapter summaries have conclusions like the following:

Species alive today are descended with changes from ancestral species living in the past. This process, whereby diverse species evolved from common ancestors, unites all organisms on earth into a single tree of life.

The uniting of all organisms on earth into a single tree of life means all life came from the same initial life form. Nothing could be clearer; the uniting of all organisms on earth into a single tree of life is the essence of **macro**evolution, change from one 'kind' to another. This conclusion is often drawn without presenting one example of a transitional fossil showing **macro**evolution. It is a classic case of bait and switch.

It is time to refer once again to the National Association of Biology Teachers statement on evolution. It states Biology should be taught in a '**pedagogically honest manner.**' There is nothing honest about using **micro**evolution to justify **macro**evolution.

How many high school or even university students have the background and experience to pick up the subtle use of **micro**evolution to justify **macro**evolution?

The above chapter summary refers to a single tree of life. Does the fossil record support a single tree of life? It most emphatically does not and here is why.

The Cambrian Explosion is the name given to a very short period of geologic time when virtually all known examples of phyla appeared suddenly and without any apparent precursors.

Phylum is the scientific name given to separate groups of creatures having similar characteristics and body plans. Some examples of phyla names are; Mollusca, the shells on the seashore and Annelida, the worms. Keep in mind these creatures are placed in different phyla because they are distinctly different from one another.

Macroevolution proponents believe a tiny life form appeared. Over time it changed and developed into the many varieties of living creatures seen today. **Macro**evolution proponents may believe this, but the fossil record does not support it. The fossil record does not show a slow progression of change from one type to another.

According to **macro**evolution theory the number of phyla should slowly increase over time as the slow process of blind random mutation and natural selection modify existing 'types.' This is not what the fossil record shows.

The fossil record shows as many as twenty different phyla appearing in the Cambrian period. Only four more phyla appear throughout the remaining fossil record. The sudden appearance of twenty phyla in the Cambrian followed by only four additional phyla in the remaining fossil record is incompatible with the concept of slow gradual change over time. It is in stark opposition to the theory of **macro**evolution.

Each form of animal life (phyla) appearing in the Cambrian required different developmental programs as well as different regulatory genes and networks. These are needed for the development of each different body plan (phyla).

Research confirms that mutational disruption to developmental programs, regulatory genes and networks eventually degrades and destroys these programs, genes, and networks. Yet, mutation is the very thing **macro**evolution depends upon to create new body plans. How can the mutational mechanism of **macro**evolution create new body plans when **research** shows mutations are lethal to them? Proponents of **macro**evolution can assume mutations produce new animal forms but solid scientific data shows otherwise.

According to **macro**evolution theory all life is descended from a single life form. If all life descended from a single life form then the first evolutionary descendent of that single life form must have looked very similar to the original form. Likewise the next evolutionary descendent should look a little less like the original but very similar to the second form, and so on.

This is the essence of the blind random mutation, natural selection mechanism of **macro**evolution. The fossil record does not show anything even close to the above scenario. If life evolved according to the above scenario it should be clearly mirrored in the fossil record. Not only does the fossil record not show the above scenario, it in fact shows the **opposite** of the above scenario.

Macroevolution requires a progression from species to phyla however; the actual fossil record shows the opposite, progression from phyla to species.

You may recall the following classification system from your high school biology: Kingdom, Phylum, Class, Order, Family, Genus and Species.

In a Darwinian **macro**evolution scenario 'the common ancestor,' the first single life form, would be designated as the first species on earth. It would be the base of Darwin's tree of life. Over time minor differences would appear among members of this first species. These differences would increase over generations until members of that species became so different they could be classed as a new species.

Eventually differences between species would increase until the species became so different they needed to be divided into groups as genera. Again, over generations genera would become so different they eventually could be grouped as families. This process would repeat itself as differences continued to grow until the additional divisions of order, class and phylum were needed to classify them.

This is the scenario one would expect from natural selection acting on mutation, a slow transition from species to phyla.

What does the fossil record show? It shows a multitude of phyla suddenly appearing fully formed. Where are the species, the genera, the families, orders and classes that should have slowly created these phyla? They are all missing, before and during the Cambrian explosion. It is in younger strata the orders, families, genera and species are observed. This is the exact opposite of the Darwinian **macro**evolution hypothesis.

Whereas **macro**evolution requires development from species to phylum, the Cambrian fossil record clearly and unambiguously shows the opposite.

The fossil record shows a variety of life forms (phyla) all arising about the same time. There is no indication any one evolved from any other forms. The forms are so different from one another; it is illogical to say any one is a precursor to any other one. Do we find precursors to these many different forms earlier in the fossil record? The answer is no. The fossil record shows a radical diversity of Cambrian body plans. It is this diversity that makes it implausible they shared a common ancestor. The idea all life is descended from an original life form exists only in the imagination of **macro**evolution proponents. It does not exist in the fossil record.

The diagram below is a stylized drawing of Darwin's tree of life. There are many variations of this diagram, however they all must incorporate the following point to be consistent with **macro**evolution. The tree must have a single original branch representing the first proposed life form.

Tree of life diagram required by **macro**evolution:

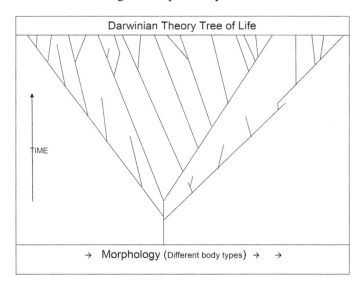

Darwinian Theory Tree of Life

TIME

→   Morphology (Different body types) →   →

The lower single beginning branch shown in the above diagram is critical to the theory of **macro**evolution. It is the essence of 'decent with modification.' In **macro**evolution theory it leads to the theoretical original common ancestor.

Careful examination of the fossil record does not support the above diagram. Diagrams of this type, beginning with a single initial vertical line, are the product of assumption, interpretation and extrapolation not the result of fossil evidence.

The following diagram is a more accurate description of the **actual** fossil record.

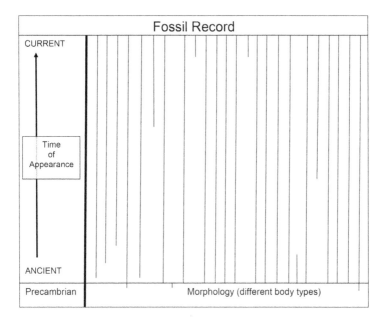

The actual fossil record depicts about twenty totally different body plans appearing at about the same time. The fossil record is quite clear; it does not show a single original branch leading to a first life form. Instead, the actual fossil record shows numerous vertical branches each quite different from the other. There is no evidence of one changing into another one.

The fossil record also shows what is called stasis. Stasis means staying the same. This is an important point.

The concept of **macro**evolution relies on a mechanism of slow continuous change. The fossil record does not show slow continuous change, it primarily shows stasis.

Some phyla became extinct while other phyla remain relatively unchanged for extremely long times. For example, sturgeon fish and horseshoe crabs existing today have undergone virtually no change throughout geologic history. If blind random mutation and natural selection are constantly at work slowly changing life, why do these and many other types seem to be immune to their effects? Stasis is the opposite of **macro**evolution. Yet stasis is the norm in the fossil record.

A solid geological principle states that in undisturbed sedimentary layers the lower layers were laid down before the layers on top of them. In strata of this type there appears to be a pattern from less complex life to more complex life from lower to higher sedimentary layers. This apparent pattern indicates less complex life appeared on the earth before the more complex life. It indicates the one is older than the other. As a result we see fossils of fish then amphibians then reptiles then mammals. Their appearance in the fossil record would seem to indicate there were fish on earth before amphibians and amphibians before reptiles and reptiles before mammals.

The above pattern is only as good as the geologic column it is based upon. In chapter eight the assumptions of the geologic column will be addressed.

The pattern described above may indicate the appearance of different kinds, however that is all it tells us. It does not tell us mammals came from reptiles and reptiles came from amphibians and amphibians came from fish. If that were so, there would be hundreds of transitional fossils that would be part of one type and part of another, including **numerous** fossils that were so transitional one could not decide which type to call them.

These hypothetical fossils do not exist in the fossil record. Proponents of **macro**evolution like to display a few hypothetical examples of what they **assume** could be transitional fossils. It is wise to remember, a few hypothetical examples cannot make up for the hundreds of transitional fossils needed to validate **macro**evolution.

Recall the quote from Charles Darwin found at the beginning of this chapter when he said,

"Why is not every geological formation and every stratum full of intermediate links?"

It is unreasonable to think all these missing fossils just disappeared. It is more reasonable to agree with the fossil experts, the paleontologists, who almost unanimously agree there are no transitional **macro**evolution fossils. The idea mammals came from reptiles, reptiles from amphibians, etc., is a product of assumption, interpretation and extrapolation, not solid scientific facts.

Undisturbed strata may indicate a progression from fish to mammals, however, Genesis, the first book of the Bible, outlines virtually the same sequence. In Genesis 1:9 land appears. In Genesis 1:11 plants appear. In Genesis 1:20 sea creatures appear. In Genesis 1:24 land animals appear and in Genesis 1:26 man appears. Genesis is not a science book and so a lot of detail is absent but, the general trend is the same, a creation starting less complex and becoming more complex.

In summary, as many as twenty phyla suddenly appeared in the Cambrian. Each appeared fully formed with no apparent precursors.

Paleontologists have little faith in **macro**evolution transitional fossils. Slow change over time from one kind to another is simply not found in the fossil record. Could it be **macro**evolution theory is wrong?

Two very prominent Harvard paleontologists, Dr. J. Gould and Dr. M. Eldredge, were convinced that for all practical purposes, transitional fossils do not exist. As a result, they came up with an alternate theory.

In their theory the concept of slow gradual change is replaced by the idea of extremely rapid change. Change so rapid that the need for transitional fossils is eliminated. A theory they call punctuated equilibrium. Their theory explains away the problem of the missing transitional fossils. Unfortunately, their theory has no scientific basis. It also has a fatal weakness as its theoretical mechanism can only account for rapid fixation of a new trait, not the origin of the trait. Origination is a necessary requirement of **macro**evolution.

Donn Rosen, a curator at the American Museum of Natural History, makes this observation,

"Darwin said that speciation occurred too slowly for us to see it. Gould and Eldredge said it occurred too quickly for us to see it. Either way we don't see it."(50)

What is one to believe? Did life evolve slowly by blind random mutation and natural selection? There is little evidence of this in the fossil record.

Did life evolve rapidly in spurts, thus leaving no transitional fossils? There is no scientific evidence for this.

Since neither of the above theories seems to be in accord with the fossil record, perhaps they are both wrong. Perhaps the whole idea of 'decent with modification' is wrong. However, even mentioning the idea that decent with modification could be wrong initiates an almost hysterical backlash. Why is this? The reason is simple. Decent with modification is essential to an atheistic view of the universe. For this reason and this reason alone it is taboo to even consider it could be wrong.

Atheists will not put the theory at risk. The hard fossil data is in direct conflict with the theory, does this mean the theory needs to be re-evaluated? Under normal scientific procedures the answer would be yes. However the ideological viewpoint trumps the scientific data. Remember, it is an ideological struggle not a scientific one.

However, there are deepening cracks in the concept of the Darwinian tree of life.

Proponents of **macro**evolution have pretty much given up trying to establish a common ancestor for all living things by examination of the fossil record. The fossils are just not there.

This being the case, they have turned to the science of genetics to try and establish what the fossil record denies. They hope by following the genetic background of a variety of creatures they can establish a genetic link to a common ancestor and thus revive Darwin's tree of life. However, genetic research seems to be destroying the tree of life not reviving it.

Researchers James Degnan and Noah Rosenberg sum up their genetic research this way,

"Many of the first studies to examine the conflicting signal of different genes have found considerable discordance across gene trees: studies of hominids, pines, cichlids, finches, grasshoppers, and fruit flies have all detected genealogical discordance so widespread that no single tree topology predominates."(51)

No single tree topology predominates!! How can there be a common ancestor if no single topological tree dominates?

Michael Syvanen's comments on his genetic studies:

"We've just annihilated the tree of life, it's not a tree anymore, and it's a different topology entirely."(52)

Eric Bapteste, an evolutionary biologist at the Pierre and Marie Currie University in Paris makes this comment,

"For a long time the holy grail was to build a tree of life A few years ago it looked as though the grail was within reach. But today the project lies in tatters, torn to pieces by an onslaught of negative evidence. Many biologists now argue the tree concept is obsolete and needs to be discarded. We have no evidence at all that the tree of life is a reality."(53}

Genetic studies rather than bolstering the Darwinian tree of life appear to be in agreement with the fossil record and its denial of the common ancestor concept.

Geneticists are using a new approach known as molecular clock analysis to try and date the appearance in time of the hypothetical common ancestor.

The actual mechanism of how the molecular clock method works is complex. It requires a good understanding of genetics beyond the scope of this book. Nevertheless, it is still possible to examine the assumptions, interpretations and extrapolations necessary to achieve the highly theoretical dates the molecular clock method proposes.

The molecular clock method is riddled with assumptions. The investigators first **assume** the existence of a common ancestor labeled A. They **assume** A to be the ancestor of two other creatures labeled B and B_1. They then **assume** to know the time period in which the **assumed** common ancestor (A) lived. They then **assume** the later creatures (B and B_1) to be descendants of the common ancestor (A). By looking at differences in similar genes between (A) and

(B) and (B$_1$), while using the above **assumptions** the investigators then make a calculation as to the mutation rate from creature (A) to creatures (B) and (B$_1$). Using this **assumed** mutation rate the investigators then **extrapolate** the **assumed** mutation rate back in time to pinpoint the proposed first original common ancestor.

The extrapolation of both the above mutation rates is based on a very shaky **assumption**. The assumption that mutation rates have been constant over hundreds of millions of years, **assuming** this amount of time was even available. This is a huge assumption. Are you seeing one assumption after another followed by interpretation and extrapolation?

Geneticists assume the existence of a common ancestor, a common ancestor the fossil record denies. When faced with the problem of finding the fossils leading to this hypothetical common ancestor, they postulate the creatures were soft bodied and thus unable to be fossilized. However, multitudes of soft bodied fossils have been found in the Cambrian strata. The soft bodied theory fails and with it fails the original assumption of a common ancestor. If you start with a tenuous assumption any results it may produce are also very tenuous.

One of the hallmarks of a good theory is consistency of results. If, when tests are run the results are consistently the same, it is a sign of a good theory. The molecular clock method miserably fails this test. It does so because it is based on faulty assumptions. Remember, **macro**evolution requires one original common ancestor by definition.

Molecular clock studies have dated the age of the hypothetical common ancestor from one million years before the Cambrian to 1.5 billion years before the Cambrian. There can only be one date. Not only do different studies give widely different dates, but different genes in the same study also give different dates. These hodgepodge results do not support the theory. Only someone totally committed to the existence of a common ancestor would consider these dates reliable. These dates are a weak attempt to prop up a failed hypothesis, namely the existence of a common ancestor. These dates are widely divergent because the assumptions of the investigators are widely divergent. Interpretations from widely divergent assumptions end up in conflict with each other. This is exactly what is observed in results of molecular clock studies.

Famous American biologist Lynn Margulis quipped:

"When evolutionary biologists use computer modeling to find out how many mutations you need to get from one species to another, it's not mathematics—it's numerology." (54)

The fossil record has failed to produce the fossils required to postulate a common ancestor to all living things. It is important to note that fossils are not assumptions; they are solid structures you can hold in your hand. They are literally hard data. In science hard data should have a greater influence towards establishing truth rather than a number of assumptions.

The molecular clock method relies on assumption upon assumption. In science assumption should never be allowed to outweigh hard data. This is exactly what is being attempted. The goal is to keep the myth of a common ancestor alive. Overlooking hard data is necessary to accomplish this end. Since there is no solid scientific evidence for a common ancestor **macro**evolution cannot be a scientific fact. Remember, it is an ideological struggle not a scientific one.

In summary:

If the slow process of blind random mutation and natural selection were the driving force of **macro**evolution the fossil record would be full of transitional fossils. It is not. If blind random mutation and natural selection were the driving force of **macro**evolution it would not be possible to classify creatures as fish, amphibians, reptiles, birds and mammals, because no such obvious distinctions would be present. The fossil record would be full of a whole host of creatures that are part this and part that and intermediate to this and intermediate to that. By the same line of logic we should not be able to classify creatures living today.

Further, if over time random mutation and natural selection have such creative powers why does the fossil record show a loss of phyla since the Cambrian? Why does it not show a host of new phyla?

The truth is the fossil record is in direct opposition to the concept of blind random mutation and natural selection as a mechanism.

Proponents of **macro**evolution attempt to give occasional examples of transitional fossils. A couple of assumed examples cannot make up for the literally hundreds of transitional **macro**evolution fossils that should be easily identifiable. Whenever you see some assumed example of a transitional fossil, refer to the opinion of the many expert paleontologists who agree that for all practical purposes there are no transitional **macro**evolution fossils.

Macroevolution cannot be a scientific fact if the fossil record fails to show slow transition and genealogical studies fail to show a coherent line of common decent.

Chapter Eight

THE SECOND LAW OF THERMODYNAMICS

"The statistical probability that organic structures and the most precisely harmonized reactions that typify living organisms would be generated by accident is zero."   Physical Chemist and Nobel Laureate, Ilya Prigogine (55)

The **Second Law of Thermodynamics**, states:

"In a natural thermodynamic process, there is an increase in the sum of the entropies of the participating systems."

The second law of thermodynamics, hereafter referred to as 2LOT, is of the one most stable laws in science.

There are a number of definitions of this law, but they all encompass the idea that the universe and everything in it, is, and always will be, moving towards a state of disorder or randomness.

This means our natural world automatically goes from order to disorder, from complexity to simplicity.

For example, a car left totally unattended will eventually break down and rust away. It will never, under any circumstances, maintain and repair itself.

One way of describing this movement from complex to simple employs the term entropy. Entropy increases as disorder increases. In the example above the more the car rusts and decays the greater its entropy becomes.

Here is a simple example of entropy. Take a deck of cards and arrange them from ace to king in each suit. By arranging the cards this way the deck has more order. Its entropy has been **decreased**. If you now take the deck and throw it in the air the order previously placed upon it would be lost and the entropy of the deck has **increased**. Increasing disorder equals increasing entropy.

Living organisms work against the 2LOT. Life organizes matter. Life takes simplicity and creates complexity.

How does the 2LOT relate to **macro**evolution? Some writers attempt to show the concept of **macro**evolution violates the 2LOT. According to **macro**evolution, the production of the first theoretical self- replicating molecule would have required taking a number of simple molecules and organizing them into an extremely complex one. A great deal of organization would have taken place. This would appear to be in violation of the 2LOT which requires the opposite, a change from complex to simple. In order for the initial formation of the highly complex molecules necessary for life to take place, life needed to progress against the 2LOT.

Macroevolution theoretically shows a progression from simple to complex, from bacteria to humans. However, many proponents of **macro**evolution deny this. Here are a few examples:

- *'All extant species are equally evolved,'* Lynn Margulis and Dorian Sagan, 1995 (56).

- *'There is no progress in evolution,'* Stephen J. Gould, 1995 (57).

- *'We all agree that there's no progress,'* Richard Dawkins, 1995 (58).

- *'The fallacy of progress,'* John Maynard Smith and Eors Szathmary, 1995 (59).

The progression from simple to complex is denied because if life is progressing from simple to complex, from being less organized to being more organized, it is progressing against the 2LOT. The idea that life progresses from simple to complex violates the 2LOT unless an organizing force sufficient to drive the change from simple to complex can be found.

The earth's sun is often proposed as the energy source capable of providing the organizing force allowing life to progress against the 2LOT. It is proposed that the sun's energy can be utilized to drive the change from simple to complex. The assumption is that the gain of energy used in the progression of life would be balanced by the energy loss of the sun. If so, the 2LOT would not be violated. This is known as the compensation idea, the idea that life, causing a decrease in entropy on earth, is compensated for by the increase of the sun's entropy. The question is, how valid is the above assumption?

Scientists acknowledge that the initial formation of high energy complex molecules necessary for life is a very improbable event. Whether one considers the earth by itself (a closed system) or as receiving solar energy (an open system) life is still a very improbable event. It is not valid to say a highly improbable event suddenly becomes probable because we have moved from a closed system to an open system. Changing systems does not affect the high improbability of these molecules forming.

Proponents of **macro**evolution believe thermal energy (sunlight) is the organizing force which allows life to work against the 2LOT. They believe sunlight provided the energy needed for the original assembly of highly complex proteins and DNA.

However, there is a second kind of entropy that needs to be addressed. This second kind of entropy is called logical entropy. Logical entropy is used in informational science, computer science and communication theory. This type of entropy is sometimes referred to as configurational entropy.

Thermal energy may be capable of doing the chemical and thermal entropy work in joining amino acids together. However it is not capable of the high information coding and sequencing portion of the configurational entropy work required to produce proteins or assemble DNA.

This simply means there is a real distinction between merely joining amino acids together and joining them together in the very specific and complex ways they are found joined in living things. The same applies to the assembly of DNA.

To make this idea more clear imagine doing this exercise. Take a box of Lego blocks and randomly connect them together. It takes little energy to do this and little to no thinking, organization or information. You end up with nothing but a random bunch of blocks joined together.

Now begin joining them together to make a Lego dinosaur. Suddenly it becomes a major task involving thinking and organization. Most of us need to spend a great deal of time pouring over instructions on how to assemble the parts. The joining of amino acids to form proteins or the assembly of DNA is not merely a matter of randomly joining the Lego blocks, it is joining them in specific ways to make a specific Lego dinosaur.

It is fairly easy to join amino acids together if you do not care about the result. This is not the case for living things. Living things are not just random collections of amino acids. Their amino acids are joined in highly complex and specific ways, as are the constituent parts of DNA. Where is the instruction manual? Sunlight may be able to provide the energy to randomly join together amino acids. However, it is not capable of providing the organizing force needed to produce the manual directing the production of DNA and the joining together of amino acids in the highly specified and complex ways necessary for life.

Morowitz, Prigogine, Salisbury, Yockey, Hoyle, Wickramasinghe, Barrow, Tipler, Copredge, Bradley, Thaxton, Schroder, and Morris are all well respected scientists who have calculated the odds of thermal energy being capable of doing the configurational work of joining amino acids into proteins and the production of DNA. They all came to a similar conclusion; *thermal energy is insufficient to provide the energy required to do the specified complex coding and sequencing found in protein molecules and the production of DNA.*

Some attempts have been made to show the respected scientists listed above are wrong in their calculations. However, these attempts do this by pointing out that since we do not know how the first life began, we are unable to make any reasonable calculations about it. This is interesting because it is an admission that no one knows how life originated. **Macro**evolution cannot be a scientific fact if no one knows how life originated.

Although the origin of life by some natural process is unknown, we can make calculations about things we do know about, the specified complexity of information bearing proteins and DNA.

Chemical and thermal processes as presently understood were not capable of producing the original highly specified complex information bearing molecules like proteins and DNA.

An organizing force sufficient to produce these large complex information bearing molecules still needs to be found. It is easy to say there are no known examples of the 2LOT being violated. However, until such an organizing force is found it appears the creation of life violates the 2LOT.

Another attempt to salvage the formation of these high information complex molecules utilizes the concept of natural selection. Do not be taken in by this approach. Natural selection requires a living reproducing entity in order for it to have any effect. The formation of the molecules in question must take place **before** natural selection can work. Natural selection cannot produce the very thing that allows it to work. The cart cannot be placed before the horse.

However, the problem is even more severe. Not only is the compensation idea of thermal input incapable of producing DNA and joining amino acids in specified and complex ways, it is also illogical.

Dr. Granville Sewell, a mathematics professor at the University of Texas, comments

"The whole idea of compensation, whether by distant or nearby events, makes no sense logically: an extremely improbable event is not rendered less improbable simply by the occurrence of 'compensating' events elsewhere. According to this reasoning the second law does not prevent scrap metal from reorganizing itself into a computer in one room as long as two computers in the next room are rusting into scrap metal-and the door is open."(60)

It is worth repeating the quote from Ilya Prigogine.

"The statistical probability that organic structures and the most precisely harmonized reactions that typify living organisms would be generated by accident, is zero." Ilya Prigogine (55)

Life appears to have developed against the 2LOT. Sunlight is not capable of driving this process. Discussions of open and closed systems and compensating events are merely attempts to avoid the obvious. Life is special, unique, and consists of more than atoms and molecules. Life's organizing ability is far more complex than a mere input of sunlight.

A faith perspective postulates a Creator as the organizing force responsible for the original formation of the highly specific complex information bearing molecules necessary for life. This perspective does not inhibit research for an alternative energy source.

Proponents of **macro**evolution will cry 'God of the gaps.' However, the best scientific minds have for years and years tried to explain the existence of proteins and DNA from a naturalistic viewpoint and have been totally unsuccessful. Presently the most reasonable explanation is, they were intelligently designed.

Nobel Prize winning Molecular Biologist, Physicist and Neuroscientist, Francis Crick acknowledges the high improbability of the origin of the molecules necessary for life.

"An honest man, armed with all the knowledge available to us now, could only state that in some sense, the origin of life appears at the moment to be almost a miracle, so many are the conditions which would have had to have been satisfied to get it going. But this should not be taken to imply that there are no good reasons to believe that it could not have started on the earth by a perfectly reasonable sequence of fairly ordinary chemical reactions. The plain fact is that the time available was too long, the many microenvironments on the earth's surface too diverse, the various chemical possibilities too numerous and our own knowledge and imagination too feeble to allow us to be able to unravel exactly how it might or might not have happened such a long time ago, especially as we have no experimental evidence from that era to check our ideas against."(61)

Crick wrote these words in 1981, over thirty-five years ago. The odds of life being anything less than a miracle are greater today than when Crick postulated it. Crick addresses the possibility that a series of fairly ordinary chemical reactions over a long period of time could produce the molecules. Today it is understood that these high energy information bearing molecules cannot be produced by ordinary chemical reactions even given extremely long time periods. A number of theories have been postulated but, Stephen Meyer in his excellent book *Signature in the Cell* has effectively discounted each one.

In summary, the origination of the highly complex information bearing molecules necessary for life seems to violate the Second Law of Thermodynamics. Sixty-five years of research into the origin of life has not provided one generally accepted hypothesis. God as Creator would resolve what appears to be a scientific impasse.

Chapter Nine

SHORT LOOK AT TIME

"Time is an Illusion"   Albert Einstein (62)

Anthropologists tell us modern man has been on earth for some 200,000 years. Recent finds in Israel may push that age back as far as 400,000 years. Modern man means us, people with similar DNA footprints and basically the same size brain. Yes, there may be some minor differences but if there were major differences they would not be classed as modern humans.

In the last 6000 years on this planet we have progressed from the wheel to walking on the moon. We are able to look back 5000 years in recorded history and find examples of brilliant minds. So with a similar DNA footprint and basically the same brain size what were we doing for the previous 194,000 years?

Is it reasonable to think our ancestors wandered around in some kind of mental stupor for 194,000 years? That is 97% of our supposed existence.

It is understandable that modern man struggled with nature for five, ten, fifteen, or maybe even twenty thousand years. But, to think modern man did nothing technological for 194,000 years is beyond comprehension. Imagine 194,000 years with very little progress and suddenly within a span of 6000 years, men are walking on the moon.

Upon verification of the Israeli excavations, anthropologists will change the supposed age of modern man from 200,000 years to 400,000 years ago. This puts modern man in a mental stupor for the first 394,000 years of his existence. Does this make any sense at all? The only reasonable explanation is that modern man has not been on earth that long.

Which brings us to the question of 'time.'

How old is the earth? I do not know, but then neither does anyone else. Any number of texts will tell you the estimated age of the earth, but not one of these age determinations is based on solid science. Attempts to date the earth are based on assumptions, interpretations and extrapolations of conditions in the distant past. No one knows if these are true or not.

In the 1800's and early1900's, rocks and geological strata including those containing fossils were dated by what could only be called scientific 'guesses'. There was simply no way of accurately determining the age of a rock or strata or a fossil in strata.

Texts and encyclopedias have statements similar to the one below.

Estimates of the absolute age of geologic events amounted to little more than inspired guesswork.

Geologists were limited to using relative dating techniques for which there was no scientific basis.

In the mid 1960's radioactive dating was undertaken in an attempt to verify the earlier scientific 'guesses.'

Amazingly many of these scientific 'guesses' appear to have been extremely accurate. For example, the Cenozoic Era theoretically lasts for 60+ million years; the radioactive dating data is very similar to the earlier scientific guess. The same applies to the Paleozoic Era; once again the radioactive dating data is amazingly close to the scientific guess.

Epochs within the Cenozoic also show amazing guesswork by those early geologists. For example the dates of the Epochs Paleocene, Eocene, Oligocene and Miocene are an uncanny match to the modern radioactive dates. This is not to say that all dates are a good match, as one goes further back into deep time the dates do diverge significantly. However for the period spanning life on earth these scientific guesses are suspiciously accurate.

The real question is how could these earlier scientific guesses have been so accurate?

Are we to assume that the originators of the time periods prior to radioactive dating were so intuitive that they somehow assigned ages by scientific guesses that turned out to be almost exact? Think about that for a minute. It is just too coincidental and frankly just too convenient.

The truth is that the dating of strata and the fossils in them is completely bogus. Their dates are nothing more than scientific guesses that cannot be verified by radioactive dating. Are there radioactive dates that verify each of these time periods? The answer is no there are not.

Out of hundreds of attempts to match radioactive dates with assumed dates only a few have matched up. This amounts to nothing more than coincidence.

Do not be confused, radioactive dating only attempts to date volcanic rocks. This in no way verifies the dates of sedimentary strata or the fossils in the strata. For example, a fossil dinosaur may be dated at 50 million years because it is in strata with a particular shell fossil that has been arbitrarily dated at 50 million years. Arbitrarily dated means the date is a scientific guess. Has that strata been verified at 50 million years by radioactive dating? The answer is no.

Geological texts often contain statements similar to the following,

Paleontology (the study of fossils) is important in the study of geology. The age of rocks may be determined by the fossils found in them.

O.K. the text states the age of rocks is based on the **fossils**.

Then, a few hundred pages later the text states,

Scientists determine when fossils were formed by finding out the age of the rocks in which they lie.

Now the text states the age of fossils is based on the **rocks**. It cannot be both. The fossils cannot date the rocks at the same time the rocks date the fossils. This is a clear case of circular reasoning.

When you are told a particular **fossil** is 50 million years old, it is because it lies in strata which has been arbitrarily dated at 50 million years.

Conversely, when you are told a **stratum** is 50 million years old it is because it contains a particular fossil that has been arbitrarily dated at 50 million years. This kind of circular reasoning is rampant in fossil and strata dating.

Below is a chart showing the supposed geologic column. It gives the impression that somewhere on earth there is a place where if you dig down far enough you will find fossils of all the creatures depicted on the chart. This is simply not true.

The Geologic Column

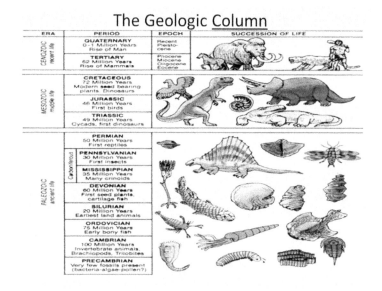

ERA	PERIOD	EPOCH	SUCCESSION OF LIFE
CENOZOIC recent life	QUATERNARY 0 - 1 Million Years Rise of Man	Recent Pleisto- cene	
	TERTIARY 62 Million Years Rise of Mammals	Pliocene Miocene Oligocene Eocene	
MESOZOIC middle life	CRETACEOUS 72 Million Years Modern seed bearing plants. Dinosaurs		
	JURASSIC 46 Million Years First birds		
	TRIASSIC 49 Million Years Cycads, first dinosaurs		
PALEOZOIC ancient life	PERMIAN 50 Million Years First reptiles		
	PENNSYLVANIAN 30 Million Years First insects		
	MISSISSIPPIAN 35 Million Years Many crinoids		
	DEVONIAN 60 Million Years First seed plants, cartilage fish		
	SILURIAN 20 Million Years Earliest land animals		
	ORDOVICIAN 75 Million Years Early bony fish		
	CAMBRIAN 100 Million Years Invertebrate animals, Brachiopods, Trilobites		
	PRECAMBRIAN Very few fossils present (bacteria-algae-pollen?)		

There is no place on earth where there is anywhere near a complete column. This chart (column) is a correlation of sedimentary layers found all over the earth. It is an assumption.

It is a fact that eighty to eighty five percent of the earth's land surface does not have even three of the ten so called geologic periods appearing in correct consecutive order as required by Darwinian evolution.

This is how a text describes the column.

The end product of correlation is a mental abstraction called the geologic column.

It then goes on to explain,

Geologists are here arguing in a circle. The succession of organisms has been determined by a study of their remains embedded in the rocks and the relative ages of the rocks are determined by the organisms they contain.

Remember, the rocks in these cases are sedimentary rocks, they cannot be dated by radioactive means.

If anyone tells you a particular rock stratum is 50 million years old because of the index fossils in the strata, ask them how they know the index fossils are 50 million years old.

If they tell you it is because of the strata the fossils are in, then ask them how they know how old the rock strata is. And round and round you go.

If they try to tell you these dates have been verified by radioactive dating ask them to show you the data, they cannot because there is no conclusive data verifying these dates.

The truth is these strata were assigned ages by scientific guesses in the 1800's and early 1900's and these ages have not been verified.

Geologist B. McKee nicely summarizes the use of scientific guesses.

"One might imagine that direct methods [radio-metric dating] of measuring time would make obsolete all of the previous means of estimating age, but these new 'absolute' measurements are used more as a supplement to traditional methods [index fossils] than as a substitute. Geologists put more faith in the principles of superposition and faunal succession than they do in numbers that come out of a machine. If the laboratory results contradict the field evidence, the geologist assumes that there is something wrong with the machine date. To put it another way, 'good' dates are those that agree with the field data." (63)

The geologic column as pictured in many geology books exists only in the mind of geologists, it does not exist in the real world. It is an assumption.

Construction of the modern time scale began in 1964 using 380 radioisotopes. These 380 radioisotopes were selected because of their agreement with the **presumed** ages of the fossil and geological sequences in rock strata. Radioisotopes not agreeing with the presumed geological sequences were rejected. In plain English, the radioisotope testing gave a range of ages for the same rock. Those involved in developing the time scale threw out the ones that did not match their presumed ages and kept the ones that did.

The same procedure continues today, albeit in a more sophisticated fashion. If a sample dates too old or too young it is subjected to a variety of adjustments. These adjustments will 'correct' the sample bringing it into line with what the samplers assumed it should be.

If these adjustments are unable to 'correct' the sample, it is discarded. If the sample agrees with the assumed date, no tests are necessary. The presupposition is all powerful.

Here is a hypothetical case to illustrate this point. From the surrounding geology (based on a manmade time scale) a particular rock sample is determined to be around 300 million years old. Radioisotope tests are run on samples of the rock. Test one dates the rock at 310 million years. It is kept as a good sample. Test two dates the rock at 420 million years. This is not in agreement with what is expected. This sample is re-worked until adjustments bring it into line with the expected age.

Other samples are subjected to the same criteria. Seemingly valid and useful rationalizations constantly bring the samples into line. If the sample cannot be 'corrected' it is discarded. It is always the sample that is wrong, never the presupposition of the samplers.

There would seem to be a simple way to test the efficiency and accuracy of radioisotope dating. Find a magma flow that is a few hundred or thousand years old that is a part of recorded history and test it. If the tests are numerous and on target then the system would seem to work. It can be matched to a known recorded date. This would also seem to have the added advantage of avoiding some of the pitfalls associated with radioisotope dating. The older the rock the greater chance that some of the daughter isotopes produced could leach out or escape through cracks in the rock. With younger rocks this should be less of a problem. Likewise testing rocks of known age should eliminate the problem of later lava flows expelling any accumulated daughter isotopes.

This type of testing has actually been done. A volcano in Hawaii is known to have erupted in 1800. Its lava is over 200 years old. A number of different radioisotope tests were run on the lava. The youngest age reported was 140 million years the oldest age was 2.96 billion years. The average age was 1.41 billion years. The actual age is just over 200 years. If we start with the average tested age of 1.41 billion years and compare it to the actual age of 200 years we find that the radioisotope dating is out by a factor of 70,500 years per year. Apply that to the current estimated age of the earth and the earth's age comes out to be about 6,400 years. There is nothing wrong with the above math, but the conclusion is rather startling.

Tests were also done on pumice from the famous Mt. Vesuvius eruption in A.D. 79. An eruption date of A.D.79 makes the pumice 1919 years old. The test results showed an age of 3,300 years. These results were then adjusted to pretty much coordinate with the correct age.

Mt. Ngauruhoe in New Zealand erupted in 1949, 1954, and 1975 making their rock samples on an average about fifty years old. They were dated at between 0.27 million years to 3.5 million years.

Mt. St. Helens erupted in 1986. Rock specimens from its lava flows dated at half a million years to 2.8 million years old.

A classic example of conflicting dates is found in the lava flows on the rim of the Grand Canyon. These lava flows are on top of all the strata in the canyon and thus must be younger than any of them. They are found in an area called the Uinkaret plateau. Human artifacts have been found in these lava flows, which mean they cannot be ancient. The following table shows a variety of dates for these lava flows.

Source	Date in years
American Indian legend	Few thousand
Stratigraphic controls	Thousands to 1 million
K – Ar radio dates	10 thousand to 3 million
Rb – Sr radio dates	0.04 to 2.6 billion

The actual age must be in the thousands of years since these volcanoes appear in Indian legends and contain human artifacts.

Once again radioactive dating appears to be unreliable when it can be measured against a known period.

Radioisotope scientists would have us believe radio-isotope dating does not work well in young age samples. However these samples are by their very nature less prone to variables that could disrupt the accuracy of the tests. These samples can be verified because they come from areas whose age is known.

These scientists would then have us believe that radioisotope dating of ancient rocks supposedly in the hundreds of millions and billions of years old is an accurate science even though the rocks are by their very nature prone to a variety of known and unknown variables. These samples cannot be verified because they come from areas whose age is unknown.

To say radioisotope dating only works well in the distant past begs the question, how do we know this? Furthermore it is logical that time would be the enemy of radioisotope testing. The older the rock is the harder it is to determine its place in history. The older the rock the more likely it may have been subjected to leaching, cracking, newer magma flows and / or processes we do not even know or understand at this time. There are conflicting results, to deny this would be dishonest. To deny that many conflicting results are adjusted, thus bringing them in line with an accepted age is also dishonest. Considering the above, it would not be prudent to be dogmatic about the accuracy of radioisotope dating.

Isochron dating is a type of radioisotope dating using a slightly different method to determine the age of a rock sample. This method was designed to eliminate the problem of being unable to determine if any daughter isotopes were present when standard testing was done. This in itself is an admission that no one knows for sure the amount of daughter isotopes present in ancient rock samples. If daughter isotopes were present or if their amount is unknown, the dating of the samples is virtually meaningless. It is an admission that radioactive dating is based on the assumption that no daughter isotopes existed when the rock was formed. This assumption is not verifiable.

However, isochron dating has its own problems. It **assumes** that all areas of a given specimen formed at the same time. It *assumes* the specimen was entirely homogenous when it formed, that no layering or mixing took place. It *assumes* no or very limited contamination took place. It *assumes* isochrones based on intra-specimen crystals can be *extrapolated* to date the whole specimen.

It *assumes* the initial slope ratios are identical at the time of eruption. Studies of young volcanic rocks at the mineral scale have shown the *assumption* that the initial slope ratios are identical at the time of eruption is not necessarily a valid assumption.

You may not understand the various technical terms used in discussing isochron dating however you can see it relies on a variety of assumptions. All radioisotope dating is heavily reliant on assumption, interpretation and extrapolation of objects and situations in the far distant past.

Radioisotope dating is not a bad idea. It is not that the theory is wrong. The mathematics is solid. These positives lead one to think the dating should be on course. The problems lie in the initial assumptions, interpretations and extrapolations the daters make. These unfortunately cannot be verified. Radioactive daters must assume:

- the rock being tested is millions of years old and

- when the rock formed there were no daughter isotopes in it, and

- over a period of millions of years any isotopes produced by radioactive decay did not leach out, or

- over a period of millions of years any isotopes produced by radioactive decay did not leach in or

- subsequent heating did not interfere with daughter isotopes levels or

- that unknown processes did not affect the isotope levels.

Recent tests have shown that radioactive elements can be easily leached out by a weak acid.

These assumptions, interpretations and extrapolations cannot be accurately checked or verified.

There is a way to check and verify the radioisotope method. A very common and very good experimental method in science is called the double blind study. In a study of this type the technicians doing the dating have no idea as to the source or presumed age of the rock being tested. Nor would they even know they were taking part in a study.

Multiple samples of rock from a variety of regions around the world, both of known age and of unknown age would be submitted for dating using a variety of radioisotope methods. The results would determine whether all methods dated samples of the same rock at the same age. It would also determine if one or more methods consistently agree more than the others. This blind study would also determine if rock dates with known ages are correctly dated by all methods. Until a comprehensive study of this type is done radioisotope dating will continue to be suspect.

It is unlikely this type of study will ever be done. A myriad of excuses would be put forth. I suspect the truth is that the experimenters already know the results would be inconclusive.

A final but very important assumption is that the rate of decay of these radioactive materials remains constant over billions of years. This assumption is based on tests measuring the rate of decay for about the last 100 years.

I defy anyone to give me another example in science where a direct extrapolation from a value of 1×10^2 to 4.5×10^8 would be considered valid science. These daters are telling us it is a fact that something they have measured for 100 years has been constant for the last at four and a half billion years.

In any other branch of science this kind of extrapolation would be laughed at. If this extrapolation is not accurate radioactive dating is useless.

Radioisotope dating is handicapped by too many assumptions, interpretations and extrapolations to be reliable. I eagerly await the results of a true double blind study which would settle the matter.

Other dating indicators such as tree rings and ice varves are used to support current theories of the earth's ancient age. However they give age ranges just in thousands of years. It is only radioactive dating that gives dates in tens of millions and billions of years.

The above information does not conclusively settle the age of the earth, but reminds us that the age of the earth is not a fixed entity. The sad fact is, only indicators which tend toward an old age for the earth are considered reliable. Many examples which would suggest a much younger age are explained away or ignored.

Another technique is Carbon-14 dating. It is supposedly reliable for about 40 to 50 thousand years into the past.

Carbon-14 dating can only be used on things that at one time were living. This is because living things take in carbon, primarily by eating food or absorbing carbon dioxide, and then at death no more carbon is taken in. A small portion of the carbon taken in is in the form of a radioactive isotope known as Carbon-14. At death the remaining Carbon-14 continues to decay and by measuring the amount of decay product and knowing the rate of decay, the time span from death to present can be fixed.

The Carbon-14 method has been plagued with variable results. It has shown some accuracy for shorter periods of time, but has also given some wildly improbable dates.

Keep in mind the radioisotope dating discussed earlier when checked with known dates was found to be unreliable.

An area of concern is the growing discrepancy between Carbon-14 dating and the radioisotope dating discussed earlier. When the same strata is dated by Carbon-14 and also indirectly dated by the radioisotope methods, the

Carbon-14 dates are in conflict with the radioisotope dates. Strata dated by both the Carbon-14 method and the radioisotope methods are found to be millions of years apart. How can the same strata be dated millions of years apart? How can Carbon-14 dates in the thousands of years be given dates in the millions of years by the radioisotope method even though the same strata is being dated?

If Carbon-14 dates an object to be thousands of years old and radioisotopes date it to be hundreds of millions of years old, both dates cannot be accurate. Some carbon-14 dates have been found to be accurate, whereas radioisotope dates tend to be all over the place when checked against known ages. It would seem at this time the Carbon-14 dates may be the more accurate. If these Carbon-14 dates are accurate it will be necessary to re-date numerous objects and strata. It is unlikely this will be done. This is a case where the data cannot be reconciled with the prevailing theory. Thus it is likely the scientific community will find some means of adjusting it or simply ignore it.

Intact blood and muscle tissue is being found in dinosaur bones which are dated at over sixty million years. Up until these finds no one and I mean no one would have had the audacity to suggest finding blood and muscle tissue was possible in a bone over sixty million years old. It is fair enough to say there is no possibility whatsoever that blood and muscle tissue could survive intact over a period of sixty million years.

There are two possibilities to account for the presence of blood and muscle tissue in bones dated over sixty million years old. The most obvious possibility is that the bones are not that old. This possibility is totally unacceptable to proponents of **macro**evolution since it destroys the whole concept of **macro**evolution.

A second possibility is that somehow, beyond our current understanding of biology, the tissue has managed to survive intact for tens of millions of years. It will take a great deal of rationalizing, but I believe rather than face the obvious time scale problem, proponents of **macro**evolution will either accept that somehow the tissue survived or they will just ignore the data. A great deal of rationalizing will be necessary, but the alternative is not acceptable to them. It is simply inconceivable to proponents of **macro**evolution that their accepted timescale could be somewhat wrong. The facts are bent to support the theory. The theory itself is not open to discussion.

Carbon-14 decays to an imperceptible and unmeasurable amount in 50,000 years. Therefore when it is detected in any sample the sample must be less than 50,000 years old. However scientists are beginning to find Carbon-14 everywhere it shouldn't be if the earth were old.

Carbon-14 is being found in coal, oil, limestone, fossil wood, graphite, natural gas, marble, dinosaur fossils, and even in supposedly billion-year-old diamonds which are particularly resistant to contamination. Either the Carbon-14 dating is wrong or these million year dates are wrong.

One counter argument is the possibility of contamination, a truly weak argument. If these Carbon-14 findings are all due to contamination then Carbon-14 dating is useless. If they are accurate then a major re-evaluation of earth age is necessary. Since ancient earth age is central to **macro**evolution theory, the rationalization police will be out in full force. It is quite possible the concept of Carbon-14 dating may be sacrificed on the altar of **macro**evolution.

The latest attack on Carbon 14 consists of the idea that minute amounts of uranium are found everywhere and are decaying and seeking out nitrogen atoms to degrade to carbon 14 thus making the presence of carbon 14 available in ancient strata.

This is a classical example of grasping at straws to maintain the ancient age hypothesis. No matter what evidence appears to support an earth age less than billions, it is somehow rationalized away.

This allows **macro**evolution proponents to go on believing their accepted timescale in spite of the data.

In chapter 12, titled **Science in the Bible**, a case is made for dinosaurs walking the earth alongside man. These new Carbon-14 dates may well usher in another instance of science confirming the Bible.

In summary, I do not know the age of the earth, but then, neither does anyone else. When the numerous assumptions and conflicting data are taken into account, it would not be prudent to be dogmatic about the age of the earth.

Chapter Ten

AN ALTERNATIVE TO MACROEVOLUTION: THE THEORY OF INTELLIGENT DESIGN

"Nothing in biology makes sense except in the light of intelligent design." W. A. Gurba

The opening line of the National Association of Biology Teachers statement on evolution begins with the quote "Nothing in biology makes sense except in the light of evolution." T. Dobzhansky made this statement in 1973, over 40 years ago. The present day outlook on evolution is very different than it was forty years ago.

Throughout the pages of this book there are many examples of modern science and technological advances undermining the basic premise of **macro**evolution. **Macro**evolution cannot maintain its privileged position in science. It is my belief that not too long in the future we will be more comfortable re-stating Dobzhansky's quote to: "Nothing in biology makes sense except in the light of intelligent design."

Perhaps the best way to introduce intelligent design is to give a definition of intelligent design.

Intelligent design is exhibited when an object or event has an extremely low probability of occurring by chance and yet matches a discernible pattern. It is necessary to explain the terms 'extremely low probability' and 'discernible pattern.'

An event having an 'extremely low probability' means the event in question could not have happened by random chance. How is this determined?

There are estimated to be 10^{80} particles in the universe.

Using current **macro**evolution theory, the universe is assumed to have been in existence for about 10^{16} seconds.

The number of interactions between the available particles in that amount of time is calculated to be 10^{43}.

If these three factors are multiplied together the maximum possibility of an event occurring once within the problistic resources of the universe is 10^{139}. Which means, the odds of the event happening are one out of 10^{139}, an incredibly small number. Just to be on the safe side, for the purposes of intelligent design, the odds have been raised to one out of 10^{150}. This number can be written like this,

1 out of 100,000,000,000,000,000,0000,000,0000,000, 000,000,000,000,000,000,000,000,000,000,000,000,0 00,000,000,000,000,000,000,000,000,000,000,000,00 0,000,000,000,000,000,000,000,000,000,000,000,000.

At these odds, the event is just not going to happen by random chance.

What is meant by a 'discernible pattern?' A discernible pattern is an event which shows specified complexity.

Specified complexity means the event is not only very complicated but also contains useful information. This is best explained by an example.

If scientists suddenly began receiving signals from space that kept repeating the first 25 prime numbers: 2, 3, 5, 7, 11 and up to 97. They would conclude the odds of this happening by random chance are so small that chance would be ruled out. They would conclude these are signals being sent by some 'alien' civilization. These signals would be a clear case of a discernible pattern giving useful information.

When 'extreme low probability' and a clear 'discernible pattern' occur in the same event, it is always an event brought into existence be some intelligence. Intelligent design is a search for events of low probability in conjunction with specified complexity.

There are three possibilities for the occurrence of an event; chance, necessity and design. If chance and necessity can be ruled out then the event in question came into being by design. Design is always related to intelligence. This process is called using the design filter. The source of the intelligence may not always be identifiable. However, being unable to identify the intelligent source does not affect the presence of the intelligence. If you found a finely crafted arrowhead you would never be able to find the maker, but you could still recognize there was an intelligence that produced it.

The first step in the design filter is to determine if an event occurred by chance. The process of eliminating chance was discussed above in the section on low probability. The next step is to determine if an event occurred by necessity.

Necessity is any event that is caused by a rule or law of any kind. For example, if you put a ball on an inclined plane it will roll down under the influence of gravity, a natural law. Gravity is the cause, not intelligent design or random chance. A salt crystal forms when a salt solution dries up, following chemical laws.

Any time a rule or law can be identified as the cause of an event, the event is happening by necessity not design or chance.

Once chance and necessity have been ruled out as the cause of an event, the only possibility left is design. The recognition of intelligent design is determined by eliminating the only other two possibilities, chance and necessity. This is referred to as the design filter.

The recognition of intelligent design is nothing new to science. It is a common practice used regularly in a variety of scientific endeavors.

For example:

Intellectual property protection

Encyclopedia Britannica took a competitor to court accusing them of copying their book and selling it as an original. In the course of the case it came out that Encyclopedia Britannica had intentionally put a fake entry into their book. A fictional character named Peter J. Carter was portrayed as an impressionist artist. When they were able to show the competitor had the same fictional character in their book, the judge ruled the odds of the competitor making up the exact same character were so high as to not have been possible, thus eliminating chance. With necessity not being at issue the only possibility left was intelligent design. The judge ruled against the competitor

Forensic Science

If car brakes were to fail resulting in the death of the driver it would be necessary to find out if they failed out of necessity. Were the parts old and worn out, thus failing by natural decay? If by natural decay, then the death is an accident caused by the laws that govern the natural decay of objects. However, if on inspection it is found the brake line has a sharp and recent cut in it, the death is no longer determined an accident. Since car brakes cannot fail by chance, and necessity has been ruled out, intelligent design is the only option left and forensic scientists look for 'who done it?'

Cryptography

During wartime when a garbled meaningless arrangement of sounds is heard, it is important to determine whether it is a random chance arrangement or whether it is a coded message. Is it random or intelligently designed? The job of the cryptographer is to decide, random or intelligent? By using the design filter intelligent design can be identified.

Archaeology

When the Rossetta stone was first found it was a meaningless series of marks on a stone tablet. No one could tell what, if anything, they meant. However every archaeologist who looked at it agreed the markings were not caused by random chance. They were able to recognize evidence of intelligent design in the patterns and arrangements of the marks.

Arson Investigation

Did the fire start by necessity, for example a pile of oily rags in a garage, or by chance from a lightning strike? If these two causes, necessity and chance, are ruled out then arson is suspected and a source of intelligence is the cause.

S.E.T.I.

Until recently the U.S government funded a project called S.E.T.I. (Search for Extra Terrestrial Intelligence). Numerous large listening discs were set up to monitor space noise. The hope was amongst all the space noise they would find a signal that was not random noise. They hoped to find a signal imparting some specified complex information. Since there is no reason to think chance or necessity could create a highly specified and complex message from space, the conclusion would be a source of intelligence. To date no signal imparting specified complex information has been found.

The recognition of intelligent design in science is nothing new. What is new is the awareness of highly specified and complex information bearing structures occurring in living things.

Intelligent design researchers are using the same criteria, chance, necessity and design, to study living things.

The theory of Intelligent Design (ID) holds that certain features of the universe and of living things are scientifically *best explained* by an intelligent cause.

The theory of ID does not speculate on *who* or *what* the intelligent cause is.

ID believes the source of the intelligence is beyond the scope of science.

The SETI project is based on the design filter. If signals are received from space giving information that is specific and complex, random chance and necessity will be ruled out and an alien intelligence will be inferred.

DNA (deoxyribonucleic acid)

It is time to use the same criteria used in the SETI project and apply them to the most highly specific and complex information system known to man. It is time to apply it to DNA.

Inside the cell of a tiny bacterium, one of the earliest known life forms, there is a vast 'library' containing the instructions for the synthesis of cellular proteins. These proteins are the essence of life. The 'library' is DNA.

DNA:

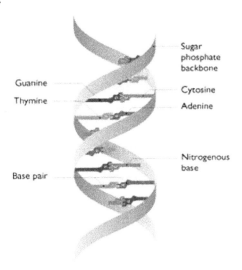

Are there any chemical or physical *laws* which determine how DNA will assemble?

DNA is made up of four nucleotides named guanine (G), adenine (A), thymine (T) and cytosine (C). There is no chemical or physical law requiring that any of the four nucleotides must come before or after any other one. Just like the letters in the alphabet they can be arranged in any order. However, just like letters in the alphabet, the nucleotides need to be assembled in a specific order to impart information. If the order of the nucleotides within DNA was determined by a chemical or physical law, it would be impossible for DNA to carry the vast amount of information necessary for life. Here is why.

When writing a sentence in English, does every letter 'a' *have* to be followed by a letter 'b'? Does some law dictate that 'e' always follow 'p'? Of course not, if there were rules that one letter had to follow another we could not spell any words and we would have no written language. Likewise there are no rules that require the nucleotides in DNA to follow one another.

DNA is a very long molecule designed to carry coded messages. In a simple cell it can be over four million 'letters' long. DNA is arranged much like books in a library. This means DNA is highly specified and very complex. It contains information that has purpose.

The design filter enables us to investigate whether the origin of DNA is due to chance, necessity or design. Remember, we are using the same criteria scientists use to identify intelligent signals from space.

Step one, chance:

What is the probability that DNA assembled by *chance* in the first cell or any cell?

It is believed that at least **300 genes** are necessary to produce a functioning organism. The statistical probability of assembling just **one** gene by chance alone is in the order of 1×10^{-190}. This means the odds of assembling 300 genes to produce a functional DNA chain by chance is essentially zero.

Chance is ruled out.

Step two, necessity:

Any nucleotide can follow any other nucleotide and that allows DNA to carry an almost infinite array of instructions. There are no laws that govern the arrangement of the nucleotides in DNA.

Necessity is ruled out.

Chance and necessity have been ruled out as originators of DNA. Once chance and necessity have been ruled out the remaining option is intelligent design. Therefore, DNA has all the hallmarks of having been designed by an intelligent source.

Earlier we learned that a meaningful sequence coming from outer space would suggest 'intelligent aliens.' This would be an acceptable scientific inference because it suggests that intelligence is the source of the signal and therefore life could exist on other planets.

DNA is a complex and meaningful 'signal' found inside all living cells. It is therefore an acceptable scientific inference that intelligence is the source for the design of DNA. Naturalistic science does not have a single generally accepted theory capable of explaining the origin of DNA.

It is interesting to note once again the atheistic bias of some scientists. Design detection is universally used in a variety of sciences however, when it points to something atheists find philosophically unacceptable, such as the arrangement of DNA nucleotides to carry information, it is no longer considered a valid means of detection.

There are numerous examples of scientific ideas that have been proposed. These ideas may flourish for some time and then either have been justified by additional research or have fallen by the wayside due to conflicting data. This is how good science works. If the design concept is a flash in the pan then future research will doom it. If it has scientific merit then future research will bolster it. What is the big deal? Why the intense animosity? Why not have a rigorous scientific debate and see who has the best explanation? Why not follow the data?

The answer to the above questions is quite simple. It is not a scientific debate. It is an ideological one. Atheists fear data showing the possibility that life is a product of design. Let's be honest, they fear there may be more to life than mere atoms and molecules.

The truth is, it is not the science they are afraid of, it is the consequence of the science that so disturbs them. If you are locked into an atheist mindset how can you tolerate a scientific line of inquiry that may destroy your ideological base?

If it were not so serious it would be amusing that these atheistic scientists, so sure they are defending science, are in fact out of tune with the general trend of modern scientific data. Their ideological bias has blinded them to the data accumulating in many different fields of science. Data is more and more reinforcing the need for design to explain many aspects of life. Science does not care about ideology.

In summary, science will slowly progress regardless of ideologies. If design is a valid requirement for many aspects of life then try as they might atheists will not be able to stifle it forever. The march of science will triumph in the end. Likewise if design is unnecessary for life it to will pass away. Let the debate begin!

Chapter Eleven

A LOVE STORY MADE IN HEAVEN

This is a love story between a male wasp, a female wasp and an orchid.

The orchid relies on the male wasp for pollination.

The female thynnid wasp is an unusual creature for two reasons. One, she cannot fly. Second, she spends virtually all of her life underground. By contrast the male wasp is quite normal. He lives above ground and flies around as you would expect a wasp to do.

This diverse lifestyle poses a few problems when it comes time to propagate the species. Somehow the two must get together. Once a year when it is time to propagate the female wasp crawls out of the ground and up a nearby stem. She then releases a powerful pheromone and waits. A pheromone is a chemical compound created by the wasp's internal chemistry which when borne on the wind is recognized by the male wasp.

The pheromone excites him to seek her out for propagation. When the male wasp picks up the scent of the pheromone he follows it to the source, the female wasp. He then picks her up and takes her to a nearby flower to feed. Somewhere along the way they mate. Having mated the male then returns her to the ground. The female goes underground for another year.

Some of you may be thinking wow, how do you explain this love story by blind random mutation and natural selection? Why make propagation so difficult, why go to all that trouble, why go underground, surely natural selection would select a simpler process? Why would blind random mutation and natural selection continue to select such complicated and extremely involved processes to propagate? While these are valid questions they are only a minor part of the story that is to follow.

The third member of this love story is the hammer orchid. The hammer orchid has somehow gotten in tune with this unusual male and female wasp arrangement. It is up to you to decide how likely it is that blind random mutation and natural selection can produce the following scenario. The fact is the hammer orchid performs as if it were aware of the mating habits of the thynnid wasp.

About the time the female wasp prepares to come above ground, the hammer orchid begins to rearrange its petals to resemble the abdomen of the female wasp. A mystery bordering on the miraculous is how blind random mutation and natural selection might re-arrange the petals of the orchid into the shape of a female wasp that lives underground. How many blind random mutations, all working towards the same end does it take to shape the orchid's petals to resemble the abdomen of the female wasp? In addition, how does blind random mutation and natural selection coordinate shaping the petals with the arrival of the female out of the ground?

Keep in mind these two actions, the re-shaping of the orchid petals and timing it with the arrival of the female wasp, must take place together to be of any value to the orchid. The shape at the wrong time would be useless. The timing without the female abdomen shape would also be useless. The odds of blind random mutations bringing about these two very different events at the same time are staggering.

Once this deceptive abdomen is complete the orchid begins to secrete the exact same pheromone that the real female wasp secretes. Not some close facsimile, the exact same chemical structure. The chemical name of this pheromone is 2 ethyl 5 propyl cyclohexane 1,3, dione. Its chemical formula looks like this:

Yes, the hammer orchid makes an exact duplicate of the wasp chemical pheromone.

The petal 'abdomen' is complete. It is available at the right time, the one and only time all year. The orchid releases the pheromone. The pheromone is essential since the male wasp follows the odor of the pheromone to find the female.

It is time to step back for a minute and see just where we are in this story.

- The orchid has prepared its fake abdomen. This in itself is of no use to the orchid. This in itself will not pollinate the orchid.

- The orchid has prepared the fake abdomen at the right time. This in itself is of no use to the orchid. This in itself will not pollinate the orchid.

- The orchid has deployed an exact replica of the female pheromone and has attracted a male wasp. This in itself is of no use to the orchid. This in itself will not pollinate the orchid.

It is important to realize that all three of the above are necessary to get both the male and female wasp to the flower. Blind random mutation and natural selection are now faced with the impossible job of bringing all three of the above changes together at the same time. Thousands of blind random mutations would be necessary to bring about these three occurrences. To believe that blind random mutation and natural selection are capable of doing this is to suspend reason.

Back to the love story, up to this point the orchid has produced a fake abdomen at the right time and has emitted an exact copy of the female pheromone.

Attracted by the pheromone, the wasp will now come and try to carry off the fake female. The wasp tries to fly away with the fake abdomen and being unsuccessful he will give up and fly away to follow another pheromone.

In doing this the wasp would not achieve the pollination needed by the flower. None of this would get the pollen dispersed. Why would the process of natural selection keep these things around? A great deal of energy is being expended for little to no return. This scenario is all but useless to the flower, all its preparation has failed to achieve its goal, pollination.

There is yet another facet to this love story. The orchid has a solution for the pollination dilemma. You see, the orchid's fake female abdomen sits on a perfectly designed lever. The lever is placed exactly the right distance from the orchid's stamens, exactly the right distance if you happen to be a male thynnid wasp. When the wasp tries to fly away with the fake female, the lever trips, bends forward and the wasp bangs his head on the orchid's stamens. He tries to fly away with the fake female a couple of more times, each time banging his head, picking up more pollen. He then gives up and flies away to follow another pheromone where he repeats the process thus pollinating another flower.

Earlier in the book we stated that life looks like it is a product of design. The complex lever apparatus is a masterpiece of design.

- The orchid constructed a fake abdomen.
- This fake abdomen appeared at just the right time.
- The orchid produced an exact duplicate of the female wasp pheromone.
- The orchid assembled a precise lever arrangement.

All of these must be in place to achieve pollination. Believing blind random mutation could produce the mutations needed to create the above arrangement is to have much more faith in blind random mutation than it takes to have faith in a Creator.

We can now see the whole system: timing, fake abdomen, pheromone, and lever need to be in place for the flower to effectively achieve pollination.

It does the orchid no good to only know when the female wasp will be coming out to mate. That alone will not get its pollen scattered.

It does the orchid no good to only create a fake abdomen. That alone will not get its pollen scattered.

It does the orchid no good to only secrete a pheromone. That alone will not get its pollen scattered.

It does the orchid no good to only construct a lever. That alone will not get its pollen scattered.

Since any single one of these cannot by itself effectively facilitate pollination they must have developed simultaneously. There would be no reason for natural selection to keep any one part for thousands of years waiting for another of the parts. Let us be very clear. The odds of blind random mutation and natural selection moving these four points along at the same time are so high as to be impossible. The only logical conclusion is something other than blind random mutation and natural selection occurs.

If you are not convinced at this point, I have saved the best for the last. You see if the male wasp is confronted with a fake abdomen and a real abdomen, both emitting the same pheromone, the male wasp can somehow pick out the real wasp from the fake. He chooses the real over the fake. This is obviously not good for the flower.

Earlier on in the story we learned that the orchid has the fake abdomen ready about the time the female wasp comes out of the ground. This is not exactly true. If this were the situation the orchid would be in direct competition with the real female wasp, a situation not to its advantage. The orchid avoids this direct competition by producing the fake abdomen, pheromone and lever system about a week to ten days before the real female wasp comes out of the ground. It then has the sole attention of the male wasp.

Proponents of **macro**evolution would have us believe that blind mutation and natural selection can randomly produce a scenario in which the orchid moves construction of the fake abdomen, pheromone production and lever construction ahead a week or ten days to avoid direct competition with the real female wasp.

It takes an enormous amount of dedication to blind random mutation and natural selection to believe they are capable of creating the above love story.

Recall the hiding and running mole story in chapter three where the professor switched the data and the evolutionists were able to come up with a satisfactory explanation for opposite occurrences. Likewise, in the above case of the wasp and the orchid, an elaborate scenario could be constructed producing a version of the story acceptable to **macro**evolution theory. However the scenario will be fabricated using copious amounts of assumption, interpretation and extrapolation.

No matter what scenario pops up, some pathway can be invented allowing the scenario to fall within the bounds of **macro**evolution theory. The theory is infinitely flexible. It can be bent to accommodate any and all data no matter how contrary the data appears.

In summary, be on the lookout for assumption, interpretation and extrapolation. Be aware of indicator words or phrases like: may have, could have, most likely, quite possibly, suggests, some think, it would seem, could be, correlation, supposition, and should be. These key words indicate assumption, interpretation and extrapolation are being used. They are used to describe theoretical ideas not factual ones. Expressions of this type can be found in abundance in any text dealing with **macro**evolution. **Macro**evolution cannot be a fact when expressions like these are used to describe it.

Chapter Twelve

SCIENCE IN THE BIBLE

"All scripture is given by the inspiration of God, and is profitable for doctrine, for reproof, for correction, for inspiration, for instruction in righteousness"   Second Timothy 3:16

The Bible is not a science book. However, whenever it intersects with science it is always right and typically years ahead of science. The following examples of Bible science are fascinating.

Genesis 17:12: "For generations to come every male among you who is eight days old must be circumcised."

A male child begins to produce vitamin K around the fifth day after birth. Vitamin K is necessary for the production of prothrombin. Prothrombin is necessary for proper blood clotting. The eighth day after birth the concentration of vitamin K in the male child rises to a level higher than it will ever be in its lifetime. Therefore, the eighth day is the best day of all to undergo a surgical procedure like circumcision which requires a healthy blood clotting system. Only God could have known this at the time Moses wrote it.

Genesis 1:1: *"In the beginning God created the heavens and the earth."*

Genesis tells us that the universe had a beginning. Up until the mid-1980s modern science insisted that the universe was eternal. That it had no beginning. That it was just always there. This was in direct contradiction to the Bible. Those believing in the Bible had to take it on faith that the universe was not eternal but had a beginning and a beginner, our God. It was only with great reluctance that many in the atheistic wing of modern science have come to agree with the Bible that the universe did indeed have a beginning.

Job 26:7: *"He spreads out the northern skies over empty space. He suspends the earth over nothing."*

Only very recently has science discovered that the earth hangs unsupported in space. For thousands of years only the Bible told us this. For example the Egyptians thought the earth was square and held up by five pillars. The Greeks said the earth was held on the back of Atlas. The Hindus said the earth was on the back of an elephant that was standing on the back of a turtle that was swimming in a cosmic sea. The above examples seem quaint and foolish to us today, but were very much believed in their time. The Bible has it right.

Job 38:31: *"Can you bind the chains of Pleiades, can you loose the cords of Orion?"*

Here we have God speaking to Job. He is telling Job something that only makes sense when viewed from the technology available to modern science. Binding Pleiades and loosing Orion; what could it mean?

Today thanks to modern technology we can at last know what it means, once again the Bible is centuries ahead of science. We now know that the stars in the constellation Pleiades are gravitationally bound together. This means they move along through space at the same rate, always the same distance apart. You can imagine this by thinking of a fighter squadron flying in formation, each keeping an equal distance from the other one yet moving along at a rapid rate. God tells Job the Pleiades are bound.

The stars in the constellation Orion are not gravitationally bound. As they move along through space they are slowly separating from each other. God asks Job, "can you loose the cords of Orion?' In the context of modern technology God's statements make perfect sense.

Job 9:8: *"He alone spreads out the heavens"*

Isaiah 44:24: *"Who stretches out the heavens all alone?"*

Zachariah 12:1: *"Thus says the Lord who stretches out the heavens"*

An Expanding Universe

In each of the above verses a stretching out of the heavens is described. The original word used is *'natah.'* It is best translated as "an effortless stretching out of one's hand". Imagine slowly raising your hand and effortlessly moving it away from your body. This represents an expansion of the heavens. Hubble described the expanding universe in 1927.

Isaiah 40:22: *"He sits enthroned above the circle of the earth"*

This verse is so clear it hardly needs explanation. The Greek word used here is *'chuwg'* (circle) and is best translated as "to encircle, to encompass or to make a circle". You will remember near the beginning of the book reading a quote by Daniel Dennett, he tried to relate non-belief in evolution to believing the earth was flat. The truth is that the Bible tells us unequivocally that the earth is round and it told us thousands of years ago long before science understood it.

Luke 17:34-36 relates the story of Jesus talking about the end time and has Him saying that two people will be in bed at night, one will be taken and the other left. Another example He relates is two women grinding grain together, one will be taken and the other left. Here we have a case of the rapture taking place at night and in the daytime. It shows an understanding of the idea that the sun is always shining somewhere and if this is the case the world must be round.

Amos 5:8*: "Seek him that maketh the seven stars and Orion,"*

This translation is from the original King James Version. The New King James version translates it thus, "He made the Pleiades and Orion," The important thing to notice is the translation of the seven stars in the original version to Pleiades in the new version. Of course there are seven stars in Pleiades. What you may not know is that when you look at the constellation Pleiades you cannot see all seven stars. You can only see six. You need a telescope to see the seventh star of Pleiades. For example in Tamil and Hindu mythology the Pleiades are referred to as the six sisters. Writings from the eighth century Japan refer to the six stars. Even today the Japanese automobile Subaru, which means

Pleiades, has an emblem on its front and rear showing six stars. There is no way Amos could have known Pleiades had seven stars by normal human knowledge. However God knew. Once again the Bible is years ahead of science.

Psalm 8:8: *"That pass through the paths of the seas."*

Mathew Fontaine Maury joined the U.S. Navy in 1825 at the age of nineteen. He served until at the age of 33 at which time an accident left him unfit for sea duty. While in hospital recovering his son entertained him reading from the book of Psalms. When his son read Psalm 8:8 Mathew Maury, who already had a love and passion for unlocking the secrets of the ocean, decided that there must literally be paths in the sea. He set out to find them. He was successful. He charted most of the major ocean currents we know today. He has been called the Father of Modern Oceanography. His belief that if the Bible said it, it must be true was the inspiration for his discoveries.

Jeremiah 33:22: *"As the host of heaven cannot be numbered,"*

The concept of an innumerable number of stars in the heavens is a very recent one. Hipparchus said in 150 B.C. that there were 1,026 stars. 150 years later Ptolemy said there were 1,056. In the early 1600's Johannes Kepler decided there were 1,006. Today, it is believed you can see somewhere between 6 and 7 thousand stars looking up from any one place on earth. Of course with the advent of the telescope and now with modern technology we understand that there are billions of stars in the heavens. The Bible tells us the host of heaven cannot be numbered. Once again the Bible was well in advance of science.

Leviticus 17:11: *"For the life of the flesh is in the blood."*

Blood carries oxygen to all our cells. It removes carbon dioxide from all our cells. It removes waste products from all our cells. Blood sustains life. It is for this reason when someone is bleeding profusely we say the life is flowing out of him. Only recently have we come to understand the life giving properties of blood.

Until recent times bloodletting was a common practice. Some think it was the excessive bloodletting George Washington received that may have led to his death. Nearly five pints of blood were drained from him. Anyone who understood that the life of the flesh is in the blood would not drain out five pints. This practice underscores that only recently have we come to understand the life giving properties of blood. The Bible recognized this centuries ago.

Job 40:15-24: Dinosaurs and man

"15 Look now at the behemoth, which I made along with you, He eats grass like an ox.

16 See now, his strength is in his hips, And his power is in his stomach muscles.

17 He moves his tail like a cedar, The sinews of his thighs are tightly knit.

18 His bones are like beams of bronze, His ribs like bars of iron.

19 He is the first of the ways of God, Only He who made him can bring near His sword.

20 Surely the mountains yield food for him And all the beasts of the field play there.

21 He is confident under the lotus trees, In a covert of reeds and marsh.

22 The lotus trees cover him with their shade, the willows by the brook surround him.

23 Indeed the river may rage , Yet he is not disturbed, He is confident, Though the Jordan rushes into his mouth,

24 Though he takes it in his eyes, or one pierces his nose with a snare.

Often we find information in the Bible hard to understand. Yet, when we finally get that understanding, we find the Bible is correct. This section of Job is one of these instances. I hypothesis the behemoth in question in verse 15 is a dinosaur.

How do I come to this hypothesis? First, it is a creature of immense strength, not just strong, but especially strong (Verse 16). Verse 17 says he moves his tail like a cedar. There are many references to the cedars of Lebanon in the Bible. They are much prized for being tall and huge and perfect for supporting structures. As a side note, some refer to the behemoth as a hippopotamus. I have seen the tail of a hippopotamus and it could not be described in any way as being like a cedar.

Moving his tail like a cedar conjures up images of a tall robust tree slowly moving back and forth in the wind. Definitely not a hippopotamus, or for that matter an elephant whose tail is miniscule.

In verse 18 it states his bones are like beams of bronze and his ribs like bars of iron. When I stand in front of one of the skeletons of a brontosaurus in a museum, I easily visualize bones of bronze and ribs of iron. No other creature I know comes close to matching this description of bones and ribs.

Verse 19 tells us that no one but God himself can bring his sword near him. Man cannot catch him or snare him. Hippopotamuses and elephants were routinely hunted and killed by man.

Finally, verse 23 states he is able to withstand a raging river without any problems. A huge dinosaur would not be disturbed by a raging river.

After some research I believe the best translation of verse 15 to be this; "Look at the behemoth which I made *along with you*". The intent of verse 15 is to convey the idea that this behemoth was alive at the time man was alive. This idea is reinforced by verse 19, for if only God can approach the behemoth with His sword; it is in effect saying that it is useless for man to approach him with a sword. In verse 24 God is saying that man cannot capture or snare him. Unless the behemoth and man are both around what would be the point of saying man cannot capture or snare him?

Finally God says in verse 15, "look at the behemoth", God expects Job to know and have a mental image of the behemoth. When you take into account the description of the behemoth and God's conversation with Job two things stand out. The description of the behemoth is best matched to that of a dinosaur and the dinosaur is alive at the time of Job.

This is not in itself such a farfetched hypothesis. Human footprints have been found in conjunction with dinosaur tracks. These footprints have so infuriated humanist atheists that one of them showed up with an iron bar and attempted to destroy one of the best prints. Atheists first said these prints were carved by man and when this was proven to be wrong they changed their attack and now say they are eroded dinosaur prints, even though five toes are visible.

Many ancient cultures made tapestries and drawings of creatures that could only be described as dinosaurs. Over eighty of these tapestries and drawings have been found to date. Most of them have shapes that we recognize as specific types of dinosaurs. We know these shapes because of reconstructions from dinosaur bones. The only way artists in these ancient cultures could have known these shapes is if they actually saw them. Atheists resist this idea because it does not agree with their timeline.

Dr. Mary Schweitzer of North Carolina State University recently isolated intact tissue and red blood cells from a dinosaur bone that is hypothetically dated at 68 million years old. It is inconceivable how anyone could entertain the idea that intact tissue and red blood cells would last for 68 million years. Let's be clear on this, it is impossible for tissue and red blood cells to survive for over 60 million years.

A more scientifically sound and reasonable explanation is the date given the dinosaur bone is wrong.

This is not the only instance of intact tissue being found in dinosaur bones. There is simply no way these bones can be millions of years old.

Reduce the age to thousands of years and dinosaurs walked with man.

Chapter Thirteen

CONCLUSION

"It is not the duty of science to defend the theory of evolution, and stick by it to the bitter end no matter which illogical and unsupported conclusions it offers. On the contrary it is expected that scientists recognize the patently obvious impossibility of Darwin's pronouncements and predictions. Let's cut the umbilical cord that tied us to Darwin for such a long time. It is choking us and holding us back."   Dr. L.L. Cohen (64)

Parents need to be fully aware of the struggle for the hearts and minds of their children. It is very real and ongoing. One linchpin of this struggle is the Biology classroom, especially Darwinian evolution. The Christian community needs to organize and actively oppose the Darwinian assumptions, interpretations and extrapolations being taught as scientific fact.

Is evolution a scientific fact?

Let's review. There are two distinct perspectives concerning evolution, the **MICRO** evolution perspective and the **MACRO** evolution perspective.

We have seen that **micro**evolution is change within a kind. We are observing **micro**evolution at work when bacteria or viruses mutate or the size of a finch beak changes or creatures lose the ability to see. **Micro**evolution consists of 'hard' science because it is observable and reproducible. The fossil record shows numerous **micro**evolution changes. **Micro**evolution is good solid science. It is important to re-emphasize these changes are 'changes within a kind.' The bacteria are still bacteria; the finches are still finches, etc. **Micro**evolution is supported with scientific facts. **Micro**evolution is evolution consistent with Genesis. It is **micro**evolution.

Macroevolution is the idea that one 'kind' can change into another 'kind.' The idea that a self-replicating molecule spontaneously appeared and over time transformed into a single cell, then into bacteria, fish, amphibians, reptiles, birds and mammals. **Macro**evolution is not 'hard' science it is historical science. It relies heavily on assumption, interpretation and extrapolation. Historical science is not observable or reproducible. **Macro**evolution has many unanswered questions such as:

- How did life begin?
- How did amino acids form on the early earth?
- How did the first proteins form?
- What is the source of DNA.?
- What is the origin of the immense volume of information found in DNA?

- Where did epigenetic information originate?

- What is the source of configurational energy required to produce the highly specific and complex molecules essential for life?

- Is the mechanism of blind random mutation and natural selection capable of producing the incredible amount of new, never before existing genetic material?

- Where are the transitional fossils in the geologic record?

With all of these unanswered questions **macro**evolution cannot possibly qualify as a scientific fact.

The best interpretation of scientific facts available clearly shows that the universe had a beginning. Hundreds of years of philosophical and scientific agreement tell us that anything with a beginning must have a beginner. If the origin of the universe is the product of design, then all that follows is a product of design, including life.

Atheists cling to **macro**evolution because it does not require a designer. If the universe is a product of design then atheists are left clinging to **macro**evolution in spite of the science. If the universe is a product of design then **macro**evolution cannot be a scientific fact.

The origin of life on earth is a scientific mystery. Historically the origin of life was not a mystery. Virtually all pioneers of science and science methodology understood life was more than mere atoms and molecules. They saw the hand of God in life. Since the advent of Darwinian **macro**evolution, science has tried to solve the mystery of life from a purely random spontaneous beginning. The effort has been totally unsuccessful. As long as the origin of life by natural causes remains a mystery **macro**evolution cannot be a scientific fact.

The building blocks of life as we know it are proteins. Proteins consist of amino acids. Before life could have spontaneously arisen on earth amino acids must have been available in abundance. If there is no generally accepted solid science on how amino acids could have randomly formed on the early earth, **macro**evolution cannot be a scientific fact.

Amino acids join together to form proteins. There are many possible combinations for joining amino acids. However, life only uses left handed alpha shaped amino acids to form proteins. The arrangement of these special amino acids into chains is highly complex and very specific. The laws of chemistry and the mathematical rules of probability indicate the likelihood of amino acids randomly forming proteins is virtually zero. If there is no generally accepted solid science confirming how amino acids randomly arranged themselves into proteins, then **macro**evolution cannot be a scientific fact.

Modern science is not kind to the theory of **macro**evolution. In fact modern science has become the antagonist of **macro**evolution. The fields of genetics and microbiology are unveiling life as a highly specified and complex organism. It is becoming more and more unlikely that the simple mechanism of blind random mutation and natural selection could have created life. This newly discovered complexity can only be described using engineering terms.

Proponents of **macro**evolution continue to cling to the theory of blind random mutation and natural selection. They continue to do this in spite of the advances of science and technology. They continue to do this in spite of the fact that life appears more and more to be a product of design.

Darwin's seminal question remains unanswered. Where are the transitional fossils? Their absence remains as Darwin said, 'the most obvious and serious objection to the theory.' A theory lost in antiquity still searching for solid fossil remains to support it. Hundreds of thousands of fossils have been unearthed to date. However the numerous transitional fossils required for life to develop from molecule to man are not there. **Macro**evolution is a hypothetical idea lacking solid fossil evidence. The absence of solid fossil evidence indicates **macro**evolution cannot be a scientific fact.

Advances in science and technology have poked holes in many other aspects of **macro**evolution, a number of these can be found in Johnathon Wells book, 'Icons of Evolution.'

The origin of life appears to have proceeded against the Second Law of Thermodynamics. Life is an organizing force which is in direct opposition to the 2LOT which states that the universe and everything in it moves toward increasing disorder. Only by expenditure of energy can some system work against the 2LOT. Although the energy of sunlight may be capable of randomly joining together amino acids, sunlight appears to be insufficient to join together amino acids in the highly complex and specific ways needed to produce life. If this is the case then some new source of energy needs to be found which was capable of originally producing these highly complex and specific bonds. To date this source of energy is unknown. **Macro**evolution cannot be a scientific fact if there is no explanation capable of producing these highly complex and specific bonds.

Time is a manmade construct. **Macro**evolution requires billions of years to justify its existence. Our modern timescale was constructed by manmade estimations. Current estimations of the earth's age contain many assumptions. Recent finds relating to Carbon 14 seem to be in serious conflict with current estimations. The recent finding of intact tissue in dinosaur bones belies the age estimates of millions of years. Until these conflicts are resolved it would be wise to keep an open mind as to the actual age of the earth and thus the life on it.

Intelligent design is the wave of the future. Scientists are finding more and more examples of immensely complex life systems. Many systems are so interdependent the only way they could have come into existence is by a designing hand. The complexity of these systems goes well beyond what blind random mutation and natural selection can even hope to accomplish. Science is establishing the need for a designer.

At this point I would like to move beyond science and to offer my considered opinion. It is my opinion that **macro**evolution is a theory in decline. It is ironic that I previously thought it was just a matter of time until science answered all the questions pertaining to life. A matter of time until **macro**evolution became a scientific fact.

However, just the opposite has happened. As science has progressed the concept of **macro**evolution has come up against one stumbling block after another. Modern technology is proving to be the undoing of **macro**evolution.

Atheists have lost the scientific discussion in relation to the origin of the universe, the origin of life, the fossil record, and the mechanism of **macro**evolution. As a result they have increased the intensity of the ideological discussion which now ranges from simply ignoring the advances in science to outright scientific bullying, name calling and character assassination. Except for the atheistic bias in the halls of science, the theory of **macro**evolution would either be brushed aside or severely altered. In a world where at every turn, every system looks and appears to have been designed, the atheistic bias is resisting the obvious. However, science will triumph and the end of **macro**evolution as we know it is just a matter of time.

When someone tells you **macro**evolution is scientific fact, simply ask them to explain the unanswered **macro**evolution questions previously listed. If they attempt to answer any of the questions be on the lookout for words and phrases like; may have, it would seem, could have, most likely, quite possibly, suggests, some think, could be, correlation, supposition, and should be. These words and phrases are used to describe hypothetical ideas and cannot be used to describe scientific facts. These words and phrases indicate the use of assumption, interpretation and extrapolation, not scientific fact.

Darwin's famous book was titled 'The Origin of Species.' The truth is, nowhere in his book does Darwin document the origin of even one species. It was a purely hypothetical idea then and excluding changes within a kind, it remains so today.

Christians need to remember that science is a pure and noble undertaking which is under the control of mere humans. These humans are totally capable of being dogmatic, fallible and given to rationalization as much as anyone else. It is through this human screen that science must attempt to travel. Darwinian evolution and earth age dating are both based on a high degree of assumption, interpretation and extrapolation. Proceed with caution when investigating these areas of science, especially when they seem to directly contradict the Bible. Time is on our side. The Bible is the word of God.

The Bible is not considered a science book and was not intended to be one. However, when it brushes up against science it is always right and often years ahead of science. When we look at science pronouncements it would be wise to keep that in mind.

Paul, in his letter to the Colossians so many years ago, states:

"Beware lest anyone cheat you through philosophy and empty deceit, according to the tradition of men, according to the basic principles of the world, and not according to Christ." Colossians 2:8

St. Augustine nicely sums up current materialistic thinking.

"Thus does the world forget You, its Creator, and falls in love with what You have created instead of with You."(65)

Thank you, readers for supporting this journey toward better understanding the great creation, given to us by the great Creator, God.

God bless you!

Wayne A. Gurba

Appendix A

THE IDEOLOGICAL STRUGGLE

Christian parents are engaged in a struggle with the secular world for the hearts and minds of their children. These are our children. We have a right and a duty to oversee their upbringing and their education. We need to be aware that the struggle for the hearts and minds of our children is very real and ongoing. The following quotes will give you some idea of the magnitude of the struggle.

- "The schools cannot allow parents to influence the kind of values-education their children receive in school; that is what is wrong with those who say there is a universal system of values. Our (humanistic) goals are incompatible with theirs." Paul Haubner, Specialist for the N.E.A.

- "There is no God and there is no soul. Hence, there are no needs for the props of traditional religion. With dogma and creed excluded, immutable truth is also dead and buried. There is no room for fixed, natural laws or moral absolutes." John Dewey, Father of Modern American Public Education.

- "Parent's attitudes about what they want for their children represent one of the greatest barriers to successful implementation of school-to-work." U. S. Department of Education's Office of Educational Research and Improvement.

- "The major function of the school is the social orientation of the individual. It must seek to give him understanding of the transition to a new social order." Willard Givens, Executive Secretary, National Education Association, 1934.

- "As a matter of fact, creationism should be discriminated against... no advocate of such propaganda should be trusted to teach science classes or administer science programs anywhere or under any circumstances. Moreover if they are now doing so they should be dismissed." John Patterson, *Journal of the National Center for Science Education*, Fall, 198, page19.

- "Education is the most powerful ally of Humanism, and every American public school is a school of Humanism. What can the theistic Sunday school, meeting for an hour once a week, and teaching only a fraction of the children, do to stem the tide of a five-day program of humanistic teaching?" (Charles F. Potter, *Humanism: A New Religion* 1930, page 128.

Today in America we have a public school education system based on a humanistic value system. Parents need to be fully aware of this and participate in the education of their children so they are not totally lost to this humanistic value system.

Appendix B

QUOTES FAVORING A DESIGNED UNIVERSE

There appear to be only three possibilities for the existence of our universe.

One is, we are here by nothing more than blind chance. A multitude of astronomers, astrophysicists and mathematicians have done their calculations and the overwhelming consensus is that blind chance is not a viable alternative.

A second option is, we are just one of hundreds of billions of universes and we just happen to live in a universe that has the properties necessary for life as we know it. The main drawback with this option is, it is just pure fantasy. There is absolutely no scientific data to suggest it may be true. Science as we know it is limited to our universe. Once you start talking about other universes you are no longer talking science but have slipped into a realm somewhere between science fiction and metaphysics. It is an option atheists cling to since it does not require a designer. My own personal feeling is, but for the atheistic input, the multi-universe concept would not even be a consideration.

This brings us to the third option. This option is based on the facts that the universe had a beginning and in the first microseconds of its beginning the 34 constants that control the fate of the universe were not only set, but were set to such a high degree of fine tuning as to be beyond the realm of chance. The best interpretation of this data tells us we live in a designed universe. This is the only option that has solid scientific data to back it up. Occam's razor strongly favors this option.

These quotes are from Astrophysicists and Astronomers, experts in this field.

Arthur L. Schawlow, Professor of Physics at Stanford University, 1981 Nobel Prize in physics: **"It seems to me that when confronted with the marvels of life and the universe, one must ask why and not just how. The only possible answers are religious. . . . I find a need for God in the universe and in my own life."** Margenau, H. and R. A. Varghese, eds. *Cosmos, Bios, Theos: Scientists Reflect on Science, God, and the Origins of the Universe, Life, and Homo Sapiens* (Open Court Pub. Co., La Salle, IL, 1992).

Frank Tipler, Professor of Mathematical Physics: **"When I began my career as a cosmologist some twenty years ago, I was a convinced atheist. I never in my wildest dreams imagined that one day I would be writing a book purporting to show that the central claims of Judeo-Christian theology are in fact true, that these claims are straightforward deductions of the laws of physics as we now understand them. I have been forced into these conclusions by the inexorable logic of my own special branch of physics."**

Tipler, F.J. 1994. *The Physics of Immortality*. New York, Doubleday, Preface.
Note: Tipler since has actually converted to Christianity, hence his book, *The Physics of Christianity*.

Paul Davies, British astrophysicist: **"There is for me powerful evidence that there is something going on behind it all....It seems as though somebody has fine-tuned nature's numbers to make the Universe.... The impression of design is overwhelming."** Davies, P. 1988. *The Cosmic Blueprint: New Discoveries in Nature's Creative Ability to Order the Universe*. New York: Simon and Schuster, p. 203.

George Ellis, British astrophysicist: **"Amazing fine tuning occurs in the laws that make this [complexity] possible. Realization of the complexity of what is accomplished makes it very difficult not to use the word 'miraculous' without taking a stand as to the ontological status the word."** Ellis, G.F.R. 1993. The Anthropic Principle: Laws and Environments. *The Anthropic Principle*, F. Bertola and U. Curi, ed. New York, Cambridge University Press, p. 30.

Arno Penzias, Nobel Prize in physics: **"Astronomy leads us to a unique event, a universe which was created out of nothing, one with the very delicate balance needed to provide exactly the conditions required to permit life, and one which has an underlying (one might say 'supernatural') plan."** Margenau, H and R.A. Varghese, ed. 1992. *Cosmos, Bios, and Theos*. La Salle, IL, Open Court, p. 83.

George Greenstein, astronomer: **"As we survey all the evidence, the thought insistently arises that some supernatural agency–or, rather, Agency–must be involved. Is it possible that suddenly, without intending to, we have stumbled upon scientific proof of the existence of a Supreme Being? Was it God who stepped in and so providentially crafted the cosmos for our benefit?"** Greenstein, G. 1988. *The Symbiotic Universe*. New York: William Morrow, p. 27.

Arthur Eddington,astrophysicist: **"The idea of a universal mind or Logos would be, I think, a fairly plausible inference from the present state of scientific theory."** Heeren, F. 1995. *Show Me God*. Wheeling, IL, Searchlight Publications, p. 233.

Edward Milne,British cosmologist: **"As to the cause of the Universe, in context of expansion, that is left for the reader to insert, but our picture is incomplete without Him [God]."** Heeren, F. 1995. *Show Me God*. Wheeling, IL, Searchlight Publications, p. 166-167.

Alexander Polyakov, Soviet mathematician: **"We know that nature is described by the best of all possible mathematics because God created it."** Gannes, S. October 13, 1986. *Fortune*. p. 57

Barry Parker, cosmologist: **"Who created these laws? There is no question but that a God will always be needed."** Heeren, F. 1995. *Show Me God*. Wheeling, IL, Searchlight Publications, p. 223.

Wernher von Braun Pioneer rocket engineer **"I find it as difficult to understand a scientist who does not acknowledge the presence of a superior rationality**

behind the existence of the universe as it is to comprehend a theologian who would deny the advances of science." McIver, T. 1986. Ancient Tales and Space-Age Myths of Creationist Evangelism. *The Skeptical Inquirer* 10:258-276.

Appendix C

THE 34 CONSTANTS

Following is the list of 34 constants controlling the universe as identified by scientists. This means that the universe is not a random mess of out of control actions and explosions. What we see happening in the universe is controlled by these 34 constants. Astrophysicists tell us that if any one of the constants was even slightly different, our universe as we know it would not exist. These constants were set at their fine-tuned values in the first microseconds of the beginning of the universe. It has been shown over and over again they could not have randomly set themselves at such a high degree of fine tuning. The overwhelming question then becomes how did they get so fine-tuned? The logical answer is they were fine-tuned by intelligent design. For us as Christians that intelligence is our God.

1. Strong nuclear force constant
   if larger: no hydrogen would form; atomic nuclei for most life-essential elements would be unstable; thus, no life chemistry

if smaller: no elements heavier than hydrogen would form: again, no life chemistry.

2. Weak nuclear force constant
   if larger: too much hydrogen would convert to helium in the Big Bang; hence, stars would convert too much matter into heavy elements making life chemistry impossible
   if smaller: too little helium would be produced from the Big Bang; hence, stars would convert too little matter into heavy elements making life chemistry impossible

3. gravitational force constant
   if larger: stars would be too hot and would burn too rapidly and too unevenly for life chemistry
   if smaller: stars would be too cool to ignite nuclear fusion; thus, many of the elements needed for life chemistry would never form

4. electromagnetic force constant
   if greater: chemical bonding would be disrupted; elements more massive than boron would be unstable to fission
   if lesser: chemical bonding would be insufficient for life chemistry

5. ratio of electromagnetic force constant to gravitational force constant
   if larger: all stars would be at least 40% more massive than the sun; hence, stellar burning would be too brief and too uneven for life support
   if smaller: all stars would be at least 20% less massive than the sun, thus incapable of producing heavy elements.

6. ratio of electron to proton mass
    if larger: chemical bonding would be insufficient for life chemistry
    if smaller: same as above

7. ratio of number of protons to number of electrons
    if larger: electromagnetism would dominate gravity, preventing galaxy, star, and planet formation
    if smaller: same as above

8. expansion rate of the universe
    if larger: no galaxies would form
    if smaller: universe would collapse, even before stars formed

9. entropy level of the universe
    if larger: stars would not form within proto-galaxies
    if smaller: no proto-galaxies would form

10. mass density of the universe
    if larger: overabundance of deuterium from the Big Bang would cause stars to burn rapidly, too rapidly for life to form
    if smaller: insufficient helium from the Big Bang would result in a shortage of heavy elements

11. velocity of light
    if faster: stars would be too luminous for life support
    if slower: stars would be insufficiently luminous for life support

12. age of the universe
    if older: no solar-type stars in a stable burning phase would exist in the right (for life) part of the galaxy
    if younger: solar-type stars in a stable burning phase would not yet have formed

13. initial uniformity of radiation
    if more uniform: stars, star clusters, and galaxies would not have formed
    if less uniform: universe by now would be mostly black holes and empty space

14. average distance between galaxies
    if larger: star formation late enough in the history of the universe would be hampered by lack of material
    if smaller: gravitational tug-of-wars would destabilize the sun's orbit

15. density of galaxy cluster
    if denser: galaxy collisions and mergers would disrupt the sun's orbit
    if less dense: star formation late enough in the history of the universe would be hampered by lack of material

16. average distance between stars
    if larger: heavy element density would be too sparse for rocky planets to form
    if smaller: planetary orbits would be too unstable for life

17. fine structure constant (describing the fine-structure splitting of spectral lines) *if larger*: all stars would be at least 30% less massive than the sun
    if larger than 0.06: matter would be unstable in large magnetic fields
    if smaller: all stars would be at least 80% more massive than the sun

18. decay rate of protons
    if greater: life would be exterminated by the release of radiation

if smaller: universe would contain insufficient matter for life

19. ^{12}C to ^{16}O nuclear energy level ratio
    if larger: universe would contain insufficient oxygen for life
    if smaller: universe would contain insufficient carbon for life

20. ground state energy level for ^4He
    if larger: universe would contain insufficient carbon and oxygen for life
    if smaller: same as above

21. decay rate of ^8Be
    if slower: heavy element fusion would generate catastrophic explosions in all the stars
    if faster: no element heavier than beryllium would form; thus, no life chemistry

22. ratio of neutron mass to proton mass
    if higher: neutron decay would yield too few neutrons for the formation of many life-essential elements
    if lower: neutron decay would produce so many neutrons as to collapse all stars into neutron stars or black holes

23. initial excess of nucleons over anti-nucleons
    if greater: radiation would prohibit planet formation
    if lesser: matter would be insufficient for galaxy or star formation

24. polarity of the water molecule
    if greater: heat of fusion and vaporization would be too high for life
    if smaller: heat of fusion and vaporization would be too low for life; liquid water would not work as a

solvent for life chemistry; ice would not float, and a runaway freeze-up would result

25. supernovae eruptions
if too close, too frequent, or too late: radiation would exterminate life on the planet
if too distant, too infrequent, or too soon: heavy elements would be too sparse for rocky planets to form

26. white dwarf binaries
if too few: insufficient fluorine would exist for life chemistry
if too many: planetary orbits would be too unstable for life
if formed too soon: insufficient fluorine production
if formed too late: fluorine would arrive too late for life chemistry

27. ratio of exotic matter mass to ordinary matter mass
if larger: universe would collapse before solar-type stars could form
if smaller: no galaxies would form

28. number of effective dimensions in the early universe
if larger: quantum mechanics, gravity, and relativity could not coexist; thus, life would be impossible
if smaller: same result

29. number of effective dimensions in the present universe
if smaller: electron, planet, and star orbits would become unstable
if larger: same result

30. mass of the neutrino
if smaller: galaxy clusters, galaxies, and stars would not form

if larger: galaxy clusters and galaxies would be
too dense

31. Big Bang ripples
   if smaller: galaxies would not form; universe would
   expand too rapidly
   if larger: galaxies/galaxy clusters would be too dense
   for life; black holes would dominate; universe would
   collapse before life-site could form

32. size of the relativistic dilation factor
   if smaller: certain life-essential chemical reactions
   will not function properly
   if larger: same result

33. uncertainty magnitude in the Heisenberg uncertainty
   principle
   if smaller: oxygen transport to body cells would be
   too small and certain life-essential elements would be
   unstable
   if larger: oxygen transport to body cells would be
   too great and certain life-essential elements would be
   unstable

34. cosmological constant
   if larger: universe would expand too quickly to form
   solar-type stars

The list above was taken from Hugh Ross, "Design Evidences in the Cosmos (1989)," Reasons to Believe, January 1, 1998, table 1, http://www.reasons.org/articles/ design-evidences-in-the-cosmos-1998. Reprinted with permission from Reasons to Believe, 818 S. Oak Park Rd., Covina, CA 91724.

Appendix D

A SIMPLE PROTEIN FORMING DEMONSTRATION

The formation of proteins is a complex and chemically mystifying process. Amino acids combine to form proteins in ways that seem to go against chemical laws and probabilities. In a random mixture of amino acids there are eight possible and chemically viable combinations.

Why do these amino acids only combine in left hand alpha combinations when any of the other combinations are possible and available? No one knows the answer to this. One chemist declared, "God must have been left handed." Even more mystifying is the order in which these amino acids join together. In some cases over a thousand of them must line up in exactly the precise order for the protein to be viable.

Anyone can do the following experiment with a teen or adult group. It will illustrate the frustrating odds against the random formation of proteins from amino acids.

Some nomenclature is necessary:

L is for left handed form of the amino acid.
R is for the right handed form.
A is for the alpha shape an amino acid can take,
B is for beta shape,
G is for gamma shape,
D is for delta shape.

These are just names for specific shapes an amino acid can take. No one shape is preferred in ordinary chemical reactions. Only in the process of life forming proteins is a specific shape preferred. There is no chemical reason for the preference for left handed alpha shapes used in life forming proteins. It remains a chemical mystery why only left handed alpha forms are used.

The numbers 1, 2, and 3 are for the specific type or kind of amino acid required to form the protein. Amino acid 1 is different from amino acid 2 which is different from amino acid 3.

Each of these amino acids has a right hand form and a left hand form. Each of these three amino acids can be found in an alpha, beta, delta or gamma shape.

Any one of these types can technically be used in forming an amino acid. No type is chemically preferred over another type. However, amino acid protein chains capable of supporting life are only made from left handed alpha shapes since this is the only type and shape used to make proteins in life forms.

You are going to make a representation of a protein that is made of only three amino acids. Call the amino acids 1, 2 and 3. The goal is to join amino acids 1, 2, and 3 together from the pool of all 24 amino acid types and shapes of amino acids 1, 2, and 3. Again, remember that left handed alpha amino acids are the only ones that will sustain life. The average protein in life is between 150 and 250 amino acids so a three amino acid protein chain is by comparison extremely simple.

Step one: Cut out 24 cardboard squares one inch by one inch.

Step two: Label each piece as follows, until you have 24 labeled squares:

LA1, LB1, LG1, LD1:
LA2, LB2, LG2, LD2:
LA3, LB3, LG3, LD3:
RA1, RB1, RG1, RD1:
RA2, RB2, RG2, RD2:
RA3, RB3, RG3, RD3.

LA1 stands for left hand, alpha shape of amino acid 1,

LB1 stands for left hand, beta shape of amino acid 1,

LG1 stands for left hand, gamma shape of amino acid 1,

LD1 stands for left hand, delta shape of amino acid 1,

LA2 stands for left hand, alpha shape of amino acid 2, and so on.

From these 24 possibilities the goal is to make a simple 3 chain protein with the following amino acids. LA1, LA2, LA3. Note that each amino acid is left handed and alpha shaped as they must be. The 1, 2 and 3 are the different types of amino acids. The amino acids must be arranged in the specific order of 1 then 2 then 3. No other order will do. The 1, 2 and 3 must all be left handed and alpha shaped. A right handed one or any other shape will not produce a protein useful in sustaining life.

Place all 24 squares in a bag. Shake them up. Have someone draw one square out of the bag. If it is the LA1 square you can continue. If not, you have failed to make the protein and must start again.

Return the square to the bag, shake it up and try again. You must keep trying until you draw out LA1. Let's say you eventually draw out the LA1 square. You can now set this square down and it will be the first member of the three chain protein. You can now draw again. This time you must draw out the LA2 square. If the next chip you draw out is not the LA2 square you must put all squares back in the bag including the LA1 you previously drew out, and start over. You must draw out the three squares LA1, LA2 and LA3 in a row. Any variation on that order requires that you return all squares to the bag and start over.

You will probably find it will take a long time before you draw the three necessary squares in the correct order. Keep in mind that most proteins are 200 to 250 amino acids long. Imagine drawing out 200 squares in the correct order from a bag containing 1,600 squares.

This is the problem faced by those who would have us believe that amino acids assemble themselves into proteins by random chance. Science has been unable to establish a coherent explanation of how these amino acids join together in the correct shape and order.

Since scientists have no idea how something so improbable could take place, it is time to entertain the idea these proteins are intelligently designed. If it looks like a rose and smells like a rose it is most likely a rose. Proteins have all the earmarks of design.

Appendix E

QUOTES FROM PALEONTOLOGISTS

The main point to glean from these quotes is that Paleontologists, the experts in the field of fossils, are skeptical of the existence of transitional fossils.

Specifically, they agree that the fossil record shows little or no transitional fossils indicating change from one kind to another.

When someone tells you the fossil record shows **macro**evolution, they are saying that the following experts in the field of fossils are wrong. Since this is highly unlikely, you can be safe in maintaining that the fossil record does not show gradual transition from one kind to another. **Macro**evolution cannot be a scientific fact if this is so.

Steven M Stanley: "The known fossil record fails to document a single example of phyletic [gradual] evolution accomplishing a major morphologic transition and hence offers no evidence that the gradualistic model can be valid"

Macroevolution: Pattern and Process San Francisco: W. H. Freeman & Co., 1979, page 39.

Stanley: "[T]he fossil record itself provided no documentation of continuity-of gradual transitions from one kind of animal or plant to another of quite different form."

The New Evolutionary Timetable: Fossils, Genes, and the origin of species, Basic Books' Inc., Publishers' New York, 1981, page 40.

Mark Ridley: "The gradual change of fossil species has never been part of the evidence for evolution. In any case no real evolutionist, whether 'gradualist' or 'punctuationist' uses the fossil record as evidence in favor of the theory of evolution as opposed to special creation."

New scientist, Vol 90, 1981, page 830.

Stephen J. Gould: "[T]he absence of fossil evidence for intermediary stages between major transitions in organic design, indeed our inability, even in our imagination, to construct functional intermediates in many cases, has been a persistent and nagging problem for gradualist accounts of evolution.

'Is a new and general theory of evolution emerging?' in Maynard-Smith ed., Evolution Now: A Century after Darwin, W. H. Freeman & Co., 1982, page 140.

Niles Eldredge: "But the smooth transition from one form of life to another which is implied in the theory is — not borne out by the facts. The search for 'missing links' between various living creatures, like humans and apes, is probably fruitless because they probably never existed as distinct transitional types . . . but no one has yet found any evidence of such transitional

creatures. This oddity has been attributed to gaps in the fossil record which gradualists expected to fill when rook strata of the proper age had been found. In the last decade, however, geologists have found rock layers of all divisions of the last 500 million years and no transitional forms were contained in them- If it is not the fossil record which is incomplete then it must be the theory."

'Missing, Believed Nonexistent,' Manchester Guardian, The Washington Post Weekly, Vol. 119, No.22; 26 November 1978, page 1.

Eldredge and I. Tattersall: "We are faced more with a great leap of faith-that gradual, progressive adaptive change underlies the general pattern of evolutionary change we see in the rocks-than any hard evidence."

The Myths of Human Evolution, Columbia University Press, 1982, page 7.

Steven J. Gould: "The fossil record with its abrupt transitions offers no support for gradual change. All paleontologists know that the fossil record contains precious little in the way of intermediate forms, transitions between major groups are characteristically abrupt

"The Return of the Hopeful Monsters," Natural History, June-July, 1977, pages 22, 24.

Gould: "The extreme rarity of transitional forms in the fossil record persists as the trade secret of paleontology. The evolutionary trees that adorn our textbooks have data only at the tips and nodes of their branches; the rest is inference, however reasonable, not the evidence of fossils."

'Evolution's Erratic Pace,' Natural History, May 1977, page 12.

Gould: "The fundamental reason why a lot of paleontologists don't care much for gradualism is because the fossil record doesn't show gradual change and every paleontologist has known that ever since Cuvier. If you want to get around that you have to invoke the imperfection of the fossil record. Every paleontologist knows that most species don't change. That's bothersome if you are trained to believe that evolution ought to be gradual. In fact it virtually precludes your studying the very process you went into the school to study. Again, because you don't see it, that brings terrible distress."

Sunderland, 121-122.

Stephen M. Stanley: "For more than a century biologists have portrayed the evolution of life as a gradual unfolding. Today the fossil record is forcing us to revise this conventional view."

The New Evolutionary Timetable: Fossils, Genes, and the origin of species, Basic Books' Inc., Publishers' New York, 1981, page 3.

Stanley: "The known fossil record is not, and never has been, in accord with gradualism. what is remarkable is that' through a variety of historical circumstances, even the history of opposition has been obscured ….the majority of paleontologists felt that their evidence simply contradicted Darwin's stress on minute, slow, and cumulative changes leading to species transformation. Their story has been suppressed".

The New Evolutionary Timetable: Fossils, Genes, and the origin of species, Basic Books' Inc., Publishers' New York, 1981, page 71.

David E. Schindel: "The gradual morphological transitions between presumed ancestors and descendants, anticipated by most biologists are missing."

The Gaps in The Fossil Record, Nature, Vol 297, 27 May 1982, page 282.

Robert Barnes: "The fossil record tells us almost nothing about the evolutionary origin of phyla and classes, intermediate forms are non-existent undiscovered or not recognized."

Book review of invertebrate Beginnings in Paleobiology, 6(3), 1980, page 365.

Katherine G. Field: "There is no fossil record establishing historical continuity of structure for most characters that might be used to assess relationships among phyla."

Molecular Phylogeny of the Animal Kingdom, 'Science' Vol 239' 12 February 1988, page 7-8.

David Raup: "Darwin predicted that the fossil record should show a reasonably smooth continuum of ancestor-descendant pairs with a satisfactory number of intermediates between major groups. . . . Such smooth transitions were not found in Darwin's time. We are now more than a hundred years after Darwin and the situation is little changed. Since Darwin a tremendous expansion of paleontological knowledge has taken place" and we know much more about the fossil record than what was known in his time, but the basic situation is not much different we actually may have fewer examples of smooth transition than we had in Darwin's time because some of the old examples have turned out to be invalid when studied in more detail."

The Geological and Paleontological Arguments of Creationism," in Scientists Confront Creationism' ed' Laurie R. Godfrey Norton: New York, 1983, page 156.

NOTES

1. Charles Darwin, *Life and Letters*, 1887, Vol. 2, page 229.

2. Richard Dawkins, *The Blind Watchmaker*, 1986, page 6.

3. Charles F. Potter, *Humanism: A New Religion*, 1930, page 128.

4. J. Dunphy, *A Religion for a New Age*, The Humanist, Jan/ Feb. 1983.

5. Daniel C. Dennett, *Darwin's Dangerous Idea: Evolution and the Meanings of Life, 1995*.

6. Chester Pierce, Harvard University, *A Chronology of Education with Quotable Quotes*, The Florida Forum, Special Edition, May 1993, page 19.

7. Steven Weinberg, *At the Freedom from Religion Foundation*, San Antonio, Nov. 1999.

8. Raul O. Leguizamon, A scientific Dissent from Darwinism.

9. Richard Lewontin, *Billions and Billions of Demons*, New York review of books, Jan 1997, page 28.

10. Scott Todd in correspondence to Nature, 30 Sept. 1999.

11. Isaac Newton, *The Principa*, 1687.

12. Nicholas Copernicus, *De Revolutionibus*, 1543.

13. Thomas Aquinas, *Summa Contra Gentiles*, 1260-73, translated by Charles J. O'Neil, 1955, U. of Notre Dame Press, 1975.

14. Louis Pasteur, quoted in J.H. Tiner, *Louis Pasteur-Founder of Modern Medicine*, Mott Media, Milford, Michigan, USA, 1990, page 75.

15. Francis Bacon, *The New Atlantis, 1624*.

16. Albert Einstein quoted by Max Jammer, *Einstein and Religion*, Princeton University Press. Oct. 2002.

17. Quentin Smith, *Theism, Atheism and Big Bang Cosmology*, 1995, page 135.

18. Henry Schaefer quoted by J.L. Sheler and J.M. Schrof, *The Creation,* U.S. News & World Report, 23 Dec 1991, pages 56-64.

19. Albert Einstein quoted by Max Jammer, *Einstein and Religion*, Princeton University Press. Oct. 2002.

20. Art Battson, Professor, University of CA, Berkley, *Facts, Fossils and Philosophy*, 17 May 1997.

21. Charles Darwin, Origin of Species, 1859.

22. S. Olson, http://www.mnh.si.edu/onehundredyears/profiles/Storrs_Olson.html.

23. J.C.Fentress, Bolton Davidheiser, *Evolution and the Christian Faith*, Presbyterian & Reformed: Nutley NJ, 1969, page 194.

24. Jerry Coyne, *Nature* 412:587, 19 August 2001.

25. Freeman Dyson, *Disturbing the Universe* Harper & Row, 1979, page 250.

26. Henry Margenau and Roy Varghese, ed. 1992, *Cosmos, Bios, and Theos*, La Salle, IL, Open Court, page 52.

27. Sir Fred Hoyle, quoted in John Barrow and Frank Tipler, *The Anthropic Cosmological Principle*, Oxford University Press, 1986, page 22.

28. Owen Gingrich, *Dare a Scientist believe in Design (Evidence of Purpose),* Continuum 1994, page 25.

29. John; Maddox, *Down with the Big Bang*, Nature, 340: 425, 1989.

30. Robert Jastrow, *God and the Astronomers*, New York, W.W. Norton, 1978, page 116.

31. Robert Jastrow, Interviewed by *Christianity Today*, 6 Aug 1982.

32. Paul Davies, *God and the New Physics*, New York, Simon and Schuster, 1983, page 189.

33. Edward Harrison, *Masks of the Universe,* New York, Collier Books, 1985, page 285.

34. Alan Sandage, quoted by J.N. Willford, *Sizing up the Cosmos: An Astronomers Quest, New York Times*, 12 March 1991, page B9.

35. John O'Keefe, quoted by F. Heeren, *Show Me God*, Wheeling, IL, Searchlight Publications, 1995, page 200.

36. Luke A. Barnes, Fine Tuning of the Universe for Life, Page 58.

37. J.Y.T. Greid, ed., *The letters of David Hume,* Oxford: Clarendon Press, 1932, page, 187.

38. Bill Bryson, *A Short History of Nearly Everything,* Broadway Books, 2003 page 13.

39. Paul Davies, *The Fifth Miracle: The Search for the Origin and Meaning of Life,* 1999.

40. Bill. Bryson, *A Short History of Nearly Everything,* Broadway Books, 2003, page 287.

41. Bill. Bryson, *A Short History of Nearly Everything,* Broadway Books, 2003, page 288.

42. Garret Vanderkooi, *Evolution as a Scientific Theory,* Christianity Today, 7 May 1971, s. 13.

43. Sir Ernst B. Chain, *Medicine,* 1945.

44. Lynn Margulis, *Acquiring Genomes: A Theory Of The Origin Of Species.*

45. Charles Darwin, *Origin of Species,* Chapter 10, pages 319, 320.

46. Stephen J. Gould, *Paleobiology,* 1977, 3:147.

47. Dr. Raup, *Conflicts between Darwin and Paleontology,* Field Museum of Natural History Bulletin Jan. 1979, Vol. 50, No. 1, pages 22-29.

48. Stephen J. Gould, *Evolution's Erratic Pace,* Natural History, May 1977, page 12.

49. Miles Eldredge, *Missing, Believed Nonexistent,* Manchester Guardian (The Washington Post Weekly), Vol. 119, No. 22, 26 November 1978, page 1.

50. Donn Rosen, Quoted by Tom Bethell, *American Spectator,* September, 2013.

51. James Degnan and Noah Rosenberg, *Trends in Ecology and Evolution,* volume 24:6, March 2009.

52. Michael Syvanen, quoted by Graham Lawton, *Why Darwin Was Wrong About the Tree of Life*, New Scientist, 21 January 2009.

53. Eric Bapteste, quoted by Graham Lawton, *Why Darwin Was Wrong About the Tree of Life,* New Scientist, 21 January 2009.

54. Teresi 2011, *p*. 71 from an Interview in *Discover Magazine*.

55. Ilya Prigogine, N. Gregair, A. Babbyabtz, *Physics Today* 25, pages 23-28.

56. Lynn Margulis and Dorion Sagan, *What Is Life?* Simon and Schuster, 1995, page 44.

57. Stephen Jay Gould, interviewed in *The Third Culture*, by John Brockman, Simon and Schuster, 1995, page 52.

58. Richard Dawkins, interviewed in *The Third Culture* by John Brockman, Simon and Schuster, 1995, page 84.

59. John Maynard Smith and Eörs Szathmáry, *The Major Transitions in Evolution*, W.H. Freeman and Company Limited, 1995. p 4.

60. Dr. G. Sewell, Applied Mathematics letters, Jan. 2011.

61. Francis Crick, *Life Itself, Its Origin and Nature*, New York NY: Simon & Schuster, 1981, page 88.

62. Albert Einstein, Letter to M. Besso's widow, 25 Feb 1926.

63. McKee, B., Cascadia: *The Geologic Evolution of the Pacific Northwest*, 1972, p.25.

64. Dr. L.L. Cohen *"Darwin was Wrong" A Study in Probabilities* (1985).

65. St.Augustine of Hippo, *Confessions*, book two, 397-398 A.D

ABOUT THE AUTHOR

Wayne Gurba received his Bachelor of Science in Combined Sciences from the University of North Dakota and his Bachelor of Arts in Science Education from the University of South Florida.

Wayne taught Senior Chemistry and Biology in Australia and Papua New Guinea, where he also worked on the design and implementation of science curriculum. Moving to Florida he continued to teach Senior Chemistry and Biology for the next twenty-two years.

The first half of his thirty year science teaching career Wayne was an agnostic. He was an avid Darwinian evolutionist! He had little use for God or the Bible.

After inviting Christ into his heart, Wayne stated, "The scales fell from my eyes and the gaping inconsistencies of Darwinian evolution became apparent."

Having spent considerable time leading students away from the Bible by promoting Darwinian evolution, Wayne decided it was time to start leading them toward the Bible.

Over the next fifteen years he developed a course not only exposing these Darwinian inconsistencies, but also pointing out the trend in modern science acknowledging evidence of an intelligent designer. He is honored to say that he believes the designer is the God of the Bible and manifested in Jesus Christ.

UNDERSTANDING EVOLUTION grew out of this course. The author's intent is to shine God's scientific light on Darwinism's false doctrine. More specifically, it shows Darwinian evolution is far from scientific fact.

To honor the Great Creator, Wayne would like to see this book in the backpack of every high school student in the world.